From Stone to Star

From Stone to Star

A View of Modern Geology

Claude Allègre

Translated by Deborah Kurmes Van Dam

Harvard University Press
Cambridge, Massachusetts
London, England
1992

Library of Congress Cataloging in Publication Data
Allègre, Claude J.
[De la pièrre a l'étoile. English]
From stone to star : a view of modern geology / Claude Allègre ;
translated by Deborah Kurmes Van Dam.
p. cm.
Translation of: De la pièrre a l'étoile.
Includes bibliographical references and index.
ISBN 0-674-83866-1
1. Earth sciences—Popular works.   2. Astronomy—Popular works.
3. Geology—Popular works.   I. Title.
QE31.A4413   1992
550—dc20
91-35773
CIP

When the sky had been separated from the earth
When the earth had been separated from the sky
When the name of man had been fixed
When An had carried off the sky
When Enlil had carried off the earth . . .

The Epic of Gilgamesh

Preface

Geologists study the history of the Earth, astronomers that of the universe. The former work with hammers and compasses, the latter with telescopes. Geologists look at the ground, astronomers at the sky. For a long time these two branches of natural history remained completely separate. There was no communication between them, and thus the writing of the history of the world was fragmented. But recently this gap has been bridged and has even begun to disappear.

The examination of the atoms and nuclei of terrestrial and extra-terrestrial rocks has revealed their ages, their origins, their relationships, and their history back to and including their beginnings—their formation in the stars.

The intimate exploration of rocks has shattered the boundaries of traditional geology: its spatial limits by reaching beyond the terrestrial crust to the interior of the globe and by comparing the Earth with the other planets; its temporal limits by delving beyond fossil time to study the entire four and a half billion years of Earth's history, and even before.

The history of the world has become continuous, from the Big Bang to the appearance of Man.

This book tells part of the story of a scientific adventure I have lived that continues to this day. It therefore owes a great deal to those who accompanied or encouraged my scientific journey or whose paths have crossed mine. Daily contact with my students has also been a source of much intellectual joy and enrichment.

The book, however, would never have seen the light of day without the encouragement of Vincent Courtillot and Bernard Dupré; the help of Lydia Zerbib, Claude Mercier, and Claude Nourry; the suggestions and corrections of Jean-Paul Poirier, Jean-Louis Le Mouël, Gérard Manhès, Jean Audouze, and Claude, without whom everything would have been so much more difficult.

Preparing the English-language edition has given me the opportunity to work with two very professional people, Angela von der Lippe, science editor at Harvard University Press, and Deborah Van Dam, the translator. I thank them both. Since the publication of the French edition, Wallace Broecker's book has appeared, and its illustrations have inspired redrawings of many of the original figures.

Contents

From Stone to Star

1

The Genesis Taboo

The origin of the Earth, how it was formed and took its place among the stars in the universe, and the conditions that allowed this planet to become hospitable first to life and then to human life are questions that all human civilizations have pondered and are still asking themselves. The ways different societies approach this problem and integrate it into their systems of information and their beliefs, the scenarios that suggest themselves to their curiosity or anxiety, vary greatly, but they constitute one of the bases for the philosophical and metaphysical reflections of each civilization. The problem of the Earth's origin certainly concerns science, but it also goes far beyond it.

Geology is the scientific discipline whose object is the study of the Earth, its structure, its history, and its evolution. For about 150 years, however, geology refused to concern itself with the creation of the Earth and its earliest history. Geology books and courses were unwilling even to consider these subjects. Geology colloquia and conferences ignored them. The mere expression of these questions in geologic circles was considered unseemly and sufficed to discredit the questioner's scientific credibility.

Why this prolonged silence, this avowed revulsion toward a subject that should be at the very heart of the geologic discipline? The goal of this book is to penetrate this once "forbidden" domain and use the geologist's tools to decipher the messages inscribed in rock. But before breaking the taboo that has determined the way geology has been studied for a century and a half, isn't it natural to wonder about its nature and origin? To satisfy this curiosity we must examine its history.

Neptunians and Plutonians

Scholars as diverse as Nicolaus Steno, Leonardo da Vinci, Jean-Étienne Guettard, Georges Buffon, Peter Simon Pallas, and Horace Bénédict de Saussure accomplished pioneering work in geology, but the field as we know it was born in the United Kingdom at the end of the eighteenth century. The major preoccupation of the geologists of that time was the origin of the rocks and minerals that constitute the terrestrial crust and the manner in which these materials were assembled into vast, generally stratified rocky formations. They had observed that rocks are of various kinds, colors, and compositions, that the minerals of which they are composed vary, and that sedimentary strata are sometimes piled up horizontally and sometimes folded and faulted. How did such variety come to be? The origin of fossils, a question that had divided the scholarly world (Voltaire concluded that they were oyster shells thrown away by pilgrims going to Santiago de Compostela), was no longer a subject of debate, and everyone agreed that they were the remains of extinct animals without as yet concerning themselves with their sequence—why particular flora and fauna followed each other in rock strata. The presence of ancient marine deposits on the continents had long been noted. It was believed that they were vestiges of the biblical Flood. This interpretation, which allowed scholars to see proof of the truth of the Scriptures in geology, was the basis of the *Neptunian theory*.

Although it had been proposed in a similar form fifty years earlier by Bertrand de Maillet (1748), the paternity of the Neptunian synthesis has been attributed to Abraham Gottlob Werner (1750–1817), professor of mineralogy at Freiberg in Saxony. Werner asserted that rocks and minerals were products of water. They were formed in the great ocean that covered the entire surface of the Earth at a certain time. But not all the materials formed at the same time or in a single episode. They were laid down successively—the younger ones covering the older ones—over the course of the Earth's history. Werner distinguished five episodes in that history, each of which corresponded to the formation of characteristic materials and left those materials as a kind of signature:

- During the first period the granites, gneisses, and porphyries were laid down in warm seas.
- In the second, transitional rocks, schists, and graywackes covered the primitive granites and gneisses (and in a colder ocean lived fish whose fossil remains are found in schists).
- In the third, the sea began to retreat from the continents. Limestone, sandstone, chalk, and basalt, which Werner considered a *sedimentary* rock, were laid down. Mammals appeared on Earth.
- The fourth was characterized by the emergence of small continents on which rivers and wind, the agents of erosion and transport, acted, leading to the deposition on the seafloor of the clays, sands, and gravels produced by this erosion.
- Finally, in the fifth period, when the water had completely left the continents, an intense volcanism, which was powered by the burning of deeply buried coal formations, took place.

Werner and his disciples believed that these five stages occurred over a very short time, on the order of several tens of thousands of years at most, practically a biblical time scale.

Unlike Werner, James Hutton (1726–1797), the father of the Plutonians (Pluto was the god of fire in Greek mythology), had no official university position. As a wealthy gentleman farmer he had the leisure to devote himself to the study of nature. Starting with speculations based on his geological rambles through the countryside, he gradually built up a theory of the geologic world whose various aspects he sketched out in several works, but whose complete exposition constituted the two-volume work he published in 1795, *Theory of the Earth*.

Hutton believed that the materials of the terrestrial crust had two origins. Clearly, some rocks, such as the limestones, schists, and sandstones, were formed in submarine deposits, but these he considered *secondary rocks*. They resulted from the action of erosion, followed by transport and sedimentation, on other more important rocks, the *primary rocks*. Hutton believed that primary rocks were the result of the cooling of a *hot magma* from the Earth's interior. They were not, like secondary rocks, products of water, but products of fire. He called them *igneous rocks* and singled out basalts and granites as typical examples.

Like Werner, Hutton thought that rocks had been created over the course of geologic history, but, unlike Werner, he did not believe that history proceeded in a unique, irreversible sequence. It consisted instead of a repetition of identical *cycles* succeeding each other over time. Each cycle began with the action of fire, when incandescent magmas arose from the depths, interjecting granites and basalts into existing rocks and setting off volcanic eruptions. The heat they gave off caused the folding of the geologic beds (just as the heat of an oven wrinkles the skin of a baking apple), thus creating mountains. The hot period was followed by a cold period during which water was the principal actor. It eroded the mountain peaks, creating, transporting, and depositing sands, gravels, and clays in seas and lakes. The uplift of new mountains by the internal fire caused the water to flow back to the permanent ocean and allowed sediments to emerge and dry out (to be transformed into sedimentary rocks). Then the cycle began again. All types of rocks were formed during each cycle: igneous through the action of fire, sedimentary through the action of water.

Hutton believed that the interior fire was the driving force, the creator of the primary rocks and mountains. Water was the destructive force, the one that eroded, flattened out, and leveled to form secondary rocks. The geologic cycle rolled along inexorably under the antagonistic action of these two fundamental forces in a process that had been occurring since the dawn of time. Indefinitely repeated cycles caused these effects to accumulate and produce big effects over time.

For Werner's unidirectional process, which proceeded according to an established order and resulted in characteristic rocks at each stage, Hutton substituted a uniform *cyclic* or permanent history, in which it is difficult to discern a beginning or an end. He summed up this conception in a phrase that has lived through time: geological evidence, he said, shows "no vestige of a beginning,—no prospect of an end." Hutton challenged Werner's conception of *vectorial* time with that of *cyclical* time.

Just as Werner's and Hutton's theories were fundamentally opposed, the ways in which the two men supported their ideas were also completely different. Werner based his theory on a global analysis of geological formations. The hearts of continents—the Harz massif in Germany, the Bohemian-Moravian Highlands in Czechoslovakia, the Massif Central in France—consisted of granite and gneiss. These rock formations, which look old and baked, are covered along their sides with schists containing fossil fish. They are crowned by the horizontal limestone or clay strata of tertiary sedimentary basins, such as the Paris and Aquitaine basins. Sands and gravels at the top near the surface attest to the geological activity of young streams. Werner's synthesis thus appeared to be a faithful transcription of the geological map of Europe.

Hutton attempted to demonstrate his ideas by using precise and detailed field observations made on a completely different scale. Where Werner argued on a continental scale, Hutton used a scale of a few tens of meters—the scale of walking distance and actual observation (see Figure 1). In Scotland he had observed that the horizontal sedimentary strata were cut through by veins of granite ten meters thick. While trying to confirm this observation he discovered a large mass of granite several kilometers wide and a series of layers of sedimentary rock that seemed to cover it. The surface between the two types of rock, or contact, was stressed by a virtual spray of granite veins that penetrated the sedimentary rock layers. From this observation he concluded that the granite had been intruded *after* the deposition of the sedimentary strata and that granite therefore must certainly be a product of the Earth's interior, that is, of fire not water.

Hutton's second fundamental observation, again on a walking-distance scale, concerns what is called an *angular unconformity.* He noticed that strongly folded layers of sedimentary rock were covered by other horizontal layers. Between the deposition of the first set of layers and the second, a major event, the folding of the first set, must have taken place (see Figure 2). He also noticed that this geometrical arrangement of horizontal and vertical could be superimposed (layer one forming an angle with layer two, which itself forms one with layer three, and so on). In this he saw proof that the history of the Earth was divided into two types of episodes: calm periods in which sedimentary strata were deposited horizontally on the sea floor, and succeeding disturbed periods in which these strata were folded and broken. The seas invaded the continents during the calm periods and were ejected from them during the folding periods, these episodes alternating and forming cycles. Hutton's theory seemed to fit

Figure 1 James Hutton, the founder of modern geology, from a 1787
caricature.

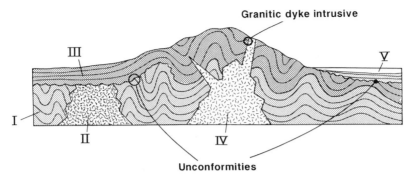

Figure 2 Diagrammatic representation of Hutton's concept of cyclic evolution. The sequence of events illustrated in the cross section from left to right:
- Sediment I deposited, then folded by a major event.
- Granite II intruded after folding.
- Erosion.
- Sediment III deposition.
- Phase of folding (including sediment III).
- Intrusion of granite IV (notice that it crosses after the later folding).
- Erosion episode.
- Deposition of sediment V.

the observed facts. Combining the geometrical relationship between granites and strata and the existing unconformity between strata, he was able to derive a local geological history from simple field observations.

Until 1790 Werner's theory was almost unanimously accepted. Like Newton's theories in physics it seemed to accord with both scientific observation and the Bible. As soon as Hutton's *Theory of the Earth* appeared, it set off a storm of controversy, since it challenged biblical truth. But Hutton himself had little chance to defend it, since he died in 1797. The battle was carried on by disciples, notably two professors at Edinburgh, John Playfair (1802) and Robert Jamieson (1808), the former for Hutton, the latter for Werner. The controversy passed rapidly beyond the geological framework of Hutton's theory and focused on its philosophical and religious implications. The Anglican church inveighed against it and, since at that time most professors of natural history were clergymen, the church could marshal considerable persuasive force.

The Flood is often cited as the major focus of this dispute, since Neptunism was concerned with the concept of successive floods while Hutton's theory was not. However important it was, I don't believe it was the only central point. The idea that the primordial geological role belonged to fire—that is, to the Devil—was also considered scandalous. According to Hutton's theory, fire, or the Devil, created the primary rocks while Heaven and the water dispensed by Heaven had only a destructive role and created only secondary rocks. The Creator or builder, therefore, was the Devil! Goethe, an occasional geologist and convinced Neptunian, was not mistaken: in his *Faust* it was Mephistopheles who defended Hutton's theories.

Even more troubling to the Christian mind was the idea that geologic time was infinite, because it suggested that infinite time, whose repeated action eventually changed all things, had more geologic power than God, who had originally created the world, that evolution dominated creation, and that Man was a latecomer in the history of the world. With Werner everything was so much simpler, so much more like the Scriptures.

Although the religious aspects of the debate were the most spectacular, the scientific arguments nevertheless continued. The Neptunians decided to refute Huttonian theory by using the same technique Hutton used: simple observation in the field. The Portrush case is a good example of those unfortunate efforts. The Irish geologist Richard Kirwan (1797) defended the Neptunian theory by claiming to have discovered a basalt containing fossils at Portrush in Northern Ireland. Basalt could not, therefore, be an igneous rock. This proclamation induced the Huttonians to go to Portrush and to demonstrate that the so-called basalt was in fact only a fossiliferous schist metamorphosed by contact with a basaltic flow.

The doom of the Wernerian thesis was sealed in a more systematic way by a series of field observations by Jean-François d'Aubuisson de Voisins (1819) and Leopold von Buch (1802), who confirmed Hutton's view on the igneous origin of granites as well as of basalts, and in the process ceased to be Wernerians and became Huttonians. The Anglican church, however, did not give up, and a pastor like the Reverend William Richardson could voice his astonishment that such a grandiose thing as the *Theory of the Earth* could be based on such a "trivial" observation as the contact between a basalt and a schist. The scientific arguments were so convincing, however, that despite the religious status of most of the professors, Hutton's theory was accepted by almost everyone in the geologic community.

Catastrophes and Uniformity of Process

In 1820 there was no indication that a new storm was brewing in the world of geology. With the triumph of Hutton's theory, geological society had become calm again and the Anglican church was at peace.

England was developing as an industrial country, which required geological engineers to lay out roads, dig canals, find coal mines, and protect soils. One of them, William Smith, gradually developed *stratigraphy,* the basic method of geology, in order to pursue his work as a civil engineer. He attempted to define, in a purely objective way, a succession of sedimentary strata, each characterized by the fossils it contained, without worrying about problems posed by these successive strata or modifications in the floral and faunal fossils (Smith, 1817). Far removed from the ideological debates, English geology was productively engaged in "serious" applied activities.

The ideological debate, however, was about to resurface. Scientific progress cannot be totally separated from theoretical interpretation, and the defeated Anglican church dreamed of a resurgence that would vindicate its old hope of "demonstrating" scientifically the authenticity of the sacred books and therefore the existence of God.

In Paris, which by 1810 had again become a world center of geological science, paleontology, and its connections with stratigraphy, was a major area of study. The savant who led this new advance was Georges Cuvier (1769–1832). Born of Swiss parents, he had been a student at Stuttgart in Germany and was now a professor at the Museum of Natural History. He first established the principles of what was to become comparative anatomy, thanks to which he was able to reconstruct the appearance of ancient animals from a few fossil remains. The discovery of an opossum in Montmartre marked the triumph of his method. Greatly influenced by the research of Johann Gottlob Lehmann and George Christian Füchsel in Saxony and Thuringia, who were strong Wernerians, Cuvier and Alexandre Brongniart, his assistant, decided to explore systematically the sedimentary strata of the Paris basin. In succeeding strata they detected a series of fossil fauna that seemed to appear suddenly and then to disappear several strata higher up. In his "Preliminary Discourse" published in 1812, Cuvier interpreted all his observations by postulating a *cyclical activity* for the entire Earth, the cycles separated by *giant catastrophes* that destroyed all living things on the continents, after which God created new species, all different from the previous ones, to replace them. One of the most striking arguments put forth by Cuvier concerned the frozen mammoths that had recently been discovered in the ice in Siberia (and can still be seen today in the Leningrad museum). If the catastrophe had not been sudden, how could these animals have allowed themselves to be frozen in place? Adolphe Brongniart, the son of Cuvier's assistant, bolstered the theory of Catastrophism by showing that the fossil flora, like the fauna, had changed nature abruptly during the course of stratigraphic succession. Clearly all living species changed over the course of time.

The catastrophist interpretation was soon extended to tectonics by Cuvier's most brilliant disciple, Léonce Elie de Beaumont (1798–1874) (Saint-Claire Deville, 1878). Using the stratigraphic method that his master had developed simultaneously with William Smith in England, and extending Hutton's deductions, he showed that the characteristic folding of mountains had been repeated over time, each mountain range being distinguished by such an episode. He placed the folding of the Pyrenees between the Cretaceous period and the Tertiary period, and that of the Alps in the Tertiary. Generalizing from these observations, Elie de Beaumont developed the idea that folding occurs during sudden periods of catastrophe, what would later be called *tectonic phases,* which he naturally linked to the observed periods of floral and faunal extinction.

The geological synthesis of the French school was certainly impressive, both in its coherence and in the multitude of observed facts it encompassed. It integrated Hutton's entire theory, notably the concepts of cyclical time, tectonic crisis, and the origins of granite and basalt as products of magmatic activity, with Werner's stratigraphic and evolutionary viewpoint, and to it joined the concept of catastrophe to explain successive natural revolutions.

Although Cuvier was a believer, he does not seem to have been very concerned about reconciling his theory with the biblical account of the Creation and the Flood, perhaps because the influence of the French Catholic church was no longer as strong as it had been in the time of the Encyclopedists fifty years earlier, and independent institutions like the Museum of Natural History did not much care about the orthodox view of the official university, the Sorbonne. For him the Flood described in the Bible was not the unique event it had been for Werner but merely one among a number of catastrophic marine invasions whose existence he had scrupulously established. Confronted with the problem of the succession of fauna, he resolved it by affirming that after each disaster, God had created a series of new species until the advent of Man.

Catastrophism was rapidly adopted and defended in the United Kingdom by a man who would become one of the great masters of English geology, William Buckland (1784–1856). A reader at Oxford who lectured even in the field attired in cap and gown, Buckland was a legendary teacher. In his inaugural lecture he announced that the object of geological research was to find the traces of biblical events and to prove the existence of God. Among the proofs of God's actions he cited the care He took to distribute coal mines in a harmonious way so that man could detect their existence from the surface. For Buckland the geological role of the Flood was not in doubt, and he adapted Cuvier's Catastrophism, gave it a religious slant, and made himself its ardent proponent. His influence was so great that he imposed his point of view on all the great English geologists: on Adam Sedgwick, the holder of the chair at Cambridge (Toulmin and Goodfield, 1965), and also on Roderick Impey Murchison, William Conybeare, and William Phillips, who drew the first geologic map of England (Gillispie, 1959; Dott and Batten, 1981). (We will come back to their work.)

Charles Lyell (1797–1875), a former student at Oxford and a disciple of Buckland, whose courses had converted him to the study of geology, was an opponent of Cuvier's Catastrophism. In 1830 he published the first volume of his classic work, *Principles of Geology,* in which he resolutely took Hutton's point of view and refuted any idea of catastrophe. He believed that all the geologic events that took place in the past and whose vestiges we see in the present were caused by phenomena identical both in nature and in intensity to those we observe today, such as erosion, sedimentation, volcanism, and earthquakes. Taking up an idea dear to Hutton, Lyell asserted that long periods of time can achieve what we, with our abbreviated lifetimes, would assume to be impossible. If the effect of time is correctly understood, it is not necessary to invoke the idea of catastrophe; it is sufficient to postulate an infinite succession of the phenomena we observe every day. Uniformity of process means that causes that are still acting in the world today are sufficient to account for the world we see around us and the vestiges of past geologic time. No recourse to catastrophe is needed.

Going even further, Lyell disputed the idea that religious beliefs about the origin of the Earth and the universe could be tested through geological observations. He believed that geology should not be confused with cosmology and that it was possible to make geology a serious science based on objective observation by studying the successions of strata in the field and fabricating rocks in the laboratory. For Lyell, the origin of the Earth (and the universe) was relevant to metaphysics, not geology. Like the Plutonians a few years earlier, the Uniformitarians would quickly win the day.

The debate between Uniformitarians and Catastrophists focused on the role of time, but the two conceptions were actually less far apart than is often thought. Both assumed the existence of repetitive cycles extending over a very long period (although for the latter, the cycles ended with a catastrophe). Neither theory afforded a glimpse of the means, or even the possibility, of studying the formation of the Earth and its earliest moments with geological methods. In this sense it could be said that both approaches were Huttonian. This convergence of views of geologic time resulted in a long-term decline in the curiosity of geologists about the Earth's formation.

Hutton's methods as elaborated by Cuvier and Lyell offered geologists a panoply of extremely useful techniques. In addition to the principle of the superposition of strata developed by the Danish geologist Nicolaus Steno (1671) in the seventeenth century, they introduced fossil chronology and the concepts of angular unconformity and magmatic intrusion. Field observations of the geometric relationships between rock formations had been translated into evolutionary schemas, and the concept of geologic cycles integrated almost all the known geologic facts.

Enthusiastically welcoming this methodological arsenal, which gave a solid scientific underpinning to their discipline, geologists at the same time accepted Lyell's uniformitarian view, with its echoes of Hutton's "no vestige of a beginning."

Without doubt the triumph of Uniformitarianism was facilitated by several factors. Industrial development called for geological engineers capable of finding coal mines or constructing canals, not for theoreticians speculating on the origin of the world. Another more religious reason seems paradoxical: the Anglican church (more than the Catholic church) had a great deal invested in geology, since the majority of English geology professors from Playfair to Buckland were members of the clergy, and the church had hoped that the rapid progress of this new science would make it possible to demonstrate the existence of God. What happened, on the contrary, was that geologic discoveries contradicted the Scriptures and challenged various points of dogma. Religious authority was not unhappy to see geology abandon a direction of research that was so inimical to it.

Uniformitarian Geology

The preceding analysis has shown how the study of Genesis was abandoned as a geologic preoccupation starting in 1850 and why geology concentrated on the study of recent periods. Now we must try to understand how this state of affairs was able to continue until perhaps 1970. Let us look first at "internal" developments in geology.

At the end of the nineteenth century geology was totally preoccupied with two major concerns. The first, which essentially concerned paleontologists and geologists working on sedimentary environments, was the controversy set off by Charles Darwin's theory (1859) of the evolution of species. The role that fossils played in these debates is well known. The debate over the origin of Man, whether expressed or implied, completely obscured that over the origin of the Earth. The second, introduced by Elie de Beaumont, was the genesis of mountains. After having worked on the origin of rocks and of nonfolded terrain and its fossils, geologists became interested in reliefs and folded areas. The Alps are young mountains, as are the Pyrenees. Geologists who were hooked on this enormous problem did not attempt to go much further back in time.

In the early twentieth century geology was soon taken up with the theory of *continental drift*. The controversy that began in 1912 when Alfred Wegener first proposed this idea preoccupied the earth sciences community from 1910 to 1930. We might recall that in Wegener's scheme, the "interesting" part of Earth's history began in the Permian, about 250 million years before the present (M.Y.B.P.), when the single supercontinent, Pangaea, began to break up, each piece drifting off on its own. The period before this did not interest Wegener any more than it did his contemporaries. An American, Thomas Chrowder Chamberlain, proposed a theory of the Earth's origin at the beginning of the twentieth century, but only astronomers were interested in it.

Prematurely abandoning Wegener's theory, geology continued to develop by returning to Hutton's idea of the *geologic cycle*. The Earth's evolution was governed by an endlessly recurring cycle: the formation of mountains associated with episodes of copious magmatism followed by the erosion of the peaks created earlier, the accumulation of sediments in geosynclinal troughs and the burying of the latter, folding due to the slow thermal contraction of the globe, and finally the uplift and formation of a new mountain range. Infinite and cyclic geologic time left no hope of deciphering the earliest history, which had been wiped out long before.

Geology in the strictest sense can thus be said to have remained profoundly uniformitarian. It must be added, however, that restricting geology to the last 500 million years offered not inconsiderable technical opportunities. During this period fossils first appear, making it possible to practice rigorously and conveniently the basic geologic method, stratigraphy. During this time also the great terrestrial reliefs that remain to this day were erected, their deep valleys providing our only three-dimensional view of tectonic structures. To go exploring in ancient rock, such as the Canadian, Brazilian, or Indian Shields, is to wander into metamorphic areas where the methods of classical geology are more difficult to apply and thus seem less convincing.

From the Second World War until 1970 the situation did not change. The emergence of the theory of *seafloor spreading* and *plate tectonics* (Holmes, 1945; Hess, 1962; Morgan, 1968), a belated resurgence of Wegener's theory, changed nothing. The majority of geologists confined themselves to what was then known to be the last 500 million years of Earth's history. Understanding what had happened during that period appeared to be the key to understanding *all* of Earth's history. Repetitive phenomenology and the cyclic idea of time had triumphed. Geology had moved away from history as evolution. Since time was cyclic, the historical perspective was useless.

If we look at the reasons for this shift, we can fit this state of affairs into the more general framework of the conflict between science and history. Physics refused to listen to Ludwig Boltzmann for a long time and confined itself to the study of equilibrium, order, simple geometric conditions, and linear systems. Astronomy, isolated for two centuries with its telescopes and its calculations of celestial mechanics, for a long time refused to consider astrophysics, which was thought to be too speculative and which was governed by dizzying space-time dimensions. This refusal in effect allowed geometry to negate history and at the same time kept astronomy out of the cosmological debate. Infinite time, as measured by the repetitive motion of the stars and planets, rejected history. Biology, whose historical character is, one could say, intrinsic or congenital, also attempted to eliminate history. Claude Bernard's experimental method introduced a new way of studying living things, not by observing long evolution and geographical dispersion as Darwin had done, but by experimenting in the laboratory. The willingness of biologists to eliminate all historical approaches was reflected in their rejection of the term *natural history* and their replacement of it first with *natural science* and later with *biology.*

It was natural that geologists, who were also seeking scientific legitimacy, would eliminate some of the "historical" content of their discipline. Synthesizing rocks in the laboratory, measuring wave propagation through the Earth, and making maps or stratigraphic cross sections were all solid, well-defined tasks whose scientific character was indisputable; and this discouraged philosophical speculation. The durability of this attitude allowed geology to become calm, serious, and rigorous. Scientific geology was constructed on the foundation of Uniformitarianism as propounded by Hutton and Lyell. No doubt this uniformitarian stage was necessary, and even useful, because it avoided the temptation of calling on mysterious, unknown, even inconceivable geologic forces at every turn. But like the paradigm of equilibrium in physics, it was only a stage in the historical development of the discipline.

Traditional Huttonian geology restricted its field of research in time and space: in time by limiting itself to recent periods; in space, by confining itself to the surface of the Earth. The emergence of a new geology was stimulated by the exploration of new areas. The determination of the internal structure of the Earth and the establishment of a calendar of geologic time would be the decisive steps in the conquest of new territories in geology, the first extending its spatial dimension, the second amplifying its time dimension.

2

Voyage to the Center

of the Earth

Surface observation makes it possible to describe horizontal or folded geological features and take samples of rocks for laboratory study over a depth of perhaps 8 kilometers, the altitude of the highest mountains. Likewise, drilling for scientific or industrial purposes can reach a depth of only 12 kilometers. Taken together, these two methods give access into the Earth of less than 20 kilometers. Since the Earth's radius is 6,400 kilometers, we can have direct knowledge of the materials of our planet for only a thin surface skin.

Everything indicates, however, that the terrestrial depths have an exciting existence. From time to time they suddenly shake the thin epidermis that constitutes the terrestrial crust and set off deadly earthquakes. They expel incandescent magmas to the surface, forming those strange glowing mountains that are called volcanoes. Their slow movements eventually displace continents and create mountain ranges. They spout hot water. From them come veins of metals and precious gems so rare and beautiful they have been the yardstick of the wealth of peoples and nations since antiquity.

The terrestrial depths, therefore, constitute a world that we would like to know, a mysterious world that appears forbidden and forever inaccessible to human observation. Just as we have come to know the atom without ever seeing or touching it, however, we can also fathom the structure of the Earth's interior without ever penetrating there ourselves.

I invite the reader on a "voyage to the center of the Earth." We will draw knowledge about our planet from this "descent into hell," which, in turn, will make it possible for us to ask fundamental questions about the Earth's origin.

Subterranean Cavities and Central Fire

For a long time it was thought that the interior of the globe was a solid pierced with holes. These cavities were believed to be of two types: some were empty or partially filled with water and linked to each other through an immense network of vast subterranean rivers and interior oceans; others were filled with hot volcanic lavas or magmas.

The relative distribution of these two types of cavities was believed to define the geologic characteristics of surface regions. Certain areas, such as southern Italy, Japan, and Iceland, were rich in magmatic pockets and therefore in volcanoes. Others, such as Greece, Yugoslavia, and Asia Minor, were famous for their subterranean caverns. The presence of the two types of subterranean cavities, hot or cold, and their alternation, combination, and eventual connection, were well described by Jules Verne in his *Journey to the Center of the Earth* published in 1864. The ideas he elaborated there are a good summary of the thinking of his time.

Superimposed on the model of a porous interior was a belief in the existence of a central fire, at least in terms of the crust. Miners of antiquity had already noticed that the deeper they dug, the hotter it got. The Earth's interior seemed to contain a source of heat, a central fire. This view, which was common toward the end of the seventeenth century, was supported by various theories, such as that of René Descartes, who hypothesized that the Earth was an aborted star. According to this philosopher, after an incandescent phase the Earth had cooled off and a superficial solid skin, the terrestrial crust, had formed. But although the cooling continued, the interior had remained somewhat soft and hot with a ball of fire in the center as a remnant of its original condition.

Not everyone believed in the idea of a central fire and the progressive cooling of the planet. Scholars as eminent as André-Marie Ampère and Siméon Denis Poisson rejected this theory, suggesting substitute hypotheses that make us smile today. But James Hutton used it to explain the creation of granites, basalts, and mountains, and it achieved fairly general acceptance. The common view of the Earth's interior was that it consisted of a ball of central fire left from the earliest days of the planet surrounded by a solid envelope that was perforated by cavities. This view persisted almost unchanged over two hundred years, from the beginning of the seventeenth to the end of the eighteenth century.

The Earth's Weight and the Enigma of the Buried Treasure

Aware of the highly theoretical character of this scheme and the absence of known facts about the physical properties of the Earth, Georges Buffon was already lamenting the fact that such models cannot be tested by measures of density: "It is known that volume for volume the Earth weighs four times as much as the Sun. We also know its weight compared to the other planets, but this is only a relative estimate, since we lack units of measurement and the real weight of matter is unknown to us: so the Earth's interior could either be empty or be filled with a material four times as heavy as gold" (Buffon, 1749).

As these remarks point out, calculating the mass of a body is a good way to determine its nature. A piece of lead is heavier than a piece of chalk; a piece of iron is heavier than a stone; water is heavier than oil. If the Earth's mass is known and its density is calculated by means of its volume, the nature of the materials that form its interior can be determined. But how can the Earth be weighed? There is no scale capable of such a feat!

The first determination of the Earth's mass was accomplished, however, in the middle of the eighteenth century (see Cook, 1970). Pierre Bourguer of France, who was sent to the Andes in 1748 to study the Earth's magnetic field, had noted that the mountains attracted his plumb line and caused it to deviate from the vertical (Bolt, 1983). This observation was used by the English astronomer Nevil Maskelyne to determine the Earth's mass. Maskelyne noticed that the deviation of the plumb line was a result of the competition between the attraction exerted by the Earth and that exerted by the mountain, the angle quantitatively reflecting their respective influences. Estimating the weight of the mountain, he deduced that of the

Earth from it. From this he calculated its density as 4.5 grams per cubic centimeter—4.5 times heavier than water. Some years later, in 1798, Lord Cavendish improved on this figure. Having measured the gravitational constant using a torsion balance, he employed Huygens's formula for the balance period of a pendulum to determine the Earth's mass. He found a density of 5.45 grams per cubic centimeter, which is nearly identical to the density of 5.25 grams per cubic centimeter accepted today.

As soon as these calculations were accepted, a fundamental fact emerged: if the rocks commonly found on the Earth's surface have densities of 2.5 to 3, the Earth as a whole is twice as heavy as its surface rocks. The interior must therefore contain a "heavy component," a region of materials whose density is much greater than that of surface rock. The density of the central core must be 7.8 to 10 grams per cubic centimeter to account for an average density of 5.4. What could this very dense element be? Copper, tin, or nickel seemed much too light. A look at the table of densities showed that lead, silver, or better yet, gold, were possible candidates. Were not gold and silver found in veins associated with rocks of deep origin, as the scientific and economic exploration of the New World had revealed? Were not Mexican and Peruvian volcanoes found in areas rich in silver or gold? The center of the Earth must thus contain an infinite reservoir of riches and deliver them up parsimoniously to the surface as a sort of metaliferous "sap" that solidified into veins.

The idea of a "buried treasure" was not unanimously accepted. For those who did not go along with the gold hypothesis, the existence of a solid or liquid center was difficult to imagine. At that time, solids and liquids were believed to be incompressible; one could not hope to increase the density of known solids by burying them. The only other possibility was the gases, which could be compressed and whose density was easily increased by increasing the pressure exerted on them. The interior of the Earth must consist of compressed gas. This theory, which in a number of ways resuscitated Descartes's ideas, was championed by Benjamin Franklin. Although the nature of the core was debated, its existence as a dense nucleus in the center of the Earth was nearly unanimously accepted.

Toward the end of the nineteenth century this hypothesis would be reinforced by one based on the properties of rotating bodies. When a spherical body turns on an axis, points situated on its equator cover a much greater distance in one revolution than points near its poles. Therefore they must turn relatively fast. Centrifugal force is greater at the equator, and the body's poles tend to flatten out or, put another way, the body "inflates" near the equator. This inflation is sizable if the mass is uniformly distributed but less important if the mass is concentrated toward the center. This property is called the *moment of inertia*. Observing the shape of the Earth and its modest inflation at the equator, where its diameter is only one three-hundredth greater than that at the poles, physicists concluded that a large part of the mass must be concentrated at the center.

Going further and using both the *moment of inertia* and the *average density,* they were able to calculate the dimensions of the dense nucleus at the center; they concluded that its radius was half that of the globe and that its density must be about 11 grams per cubic centimeter. Thus, rather simple mechanical considerations had already led to quite an elaborate model of internal structure (see Figure 3). This was the basic picture at the end of the nineteenth century. It

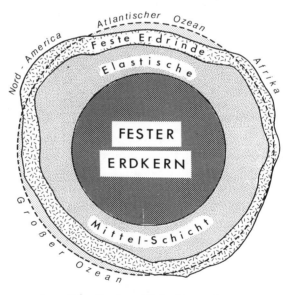

Figure 3 Cross section of the Earth published in Berlin in 1902 by H. Kraemer, before any seismological studies. Note that the Earth is shown with three layers—crust, mantle, core—even if the names are different and the proportions not quite accurate in volume.

only came to be accepted, however, as a result of a discipline that appeared in the twentieth century: seismology.

Seismology

Seismology is to the geologist what radiology is to the medical doctor. The study of the propagation of the waves emitted through the Earth by earthquakes furnishes the same information about the internal structure of the Earth as does the propagation of X rays or, more correctly, ultrasonics through the human body. Today we have seismic tomography, which is similar to medical tomography. The fact that the study of earthquakes could provide such a powerful investigative tool was realized only gradually, however.

For a long time seismology consisted of reporting, mapping, and classifying earthquakes according to the material and human destruction they caused. Then, in 1883 when he was in Japan to investigate several earthquakes that had affected that country, John Milne of England made an astonishing prediction: considering how much energy is discharged in a large earthquake, it would not be surprising if the vibrations the earthquake emits could be detected at any point on the globe.

E. von Rebeur-Paschwitz confirmed Milne's prediction six years later in Germany. He had constructed extremely precise pendulums to detect variations from the horizontal, that is, local movement in the terrain, at Pottsdam and Wilhelmshaven. On April 18, these pendulums registered extremely curious wave trains. Later he learned that on that day, a great earthquake had taken place in Tokyo. Von Rebeur-Paschwitz concluded that Tokyo was the source of the vibrations registered by his pendulums. Struck by this discovery, Richard D. Oldham of the geological survey of India began to make pendulums and record a series of large earthquakes in various parts of the globe. In 1897 he was able to formulate laws that all the recorded seismic waves seemed to obey (Oldham, 1900). Each seismogram appeared to show the arrival of two groups of small-amplitude waves later followed by large waves.

Milne used these records to study the relationship between the arrival time of each wave train and the distance separating the recording station from the source of the earthquake. He noticed that the distance could be estimated by measuring the time interval separating the arrivals of the small and large waves on the seismograms. From that it was easy to locate the earthquake by spherical triangulation using three stations (Bolt, 1983). This method, which is still in use, made it possible to map earthquakes worldwide, without ever venturing outside the laboratory, simply by processing information from at least three observations (see Figure 4).

In 1900, still pursuing his patient research on the meaning of the seismogram, Oldham demonstrated the physical interpretation of his observations. Precisely analyzing the transit times of the small waves that arrive at a station first, which he called *P* and *S* (primary and secondary), he showed that they travel through the interior of the Earth, while the large waves that arrive later travel along the surface (and are called surface waves).

The identification of the trajectories followed by the different types of seismic waves really marks the beginning of the use of seismology to determine the internal structure of the Earth. In 1906, using his understanding of the types of seismic waves, Oldham discovered the existence of two entities in the Earth's interior: in the center, the *core,* with a radius about 0.4 that of the terrestrial radius (2,550 kilometers), and the *mantle* surrounding it. The two areas were characterized by very different wave propagation velocities, and the mantle-core boundary by a sudden discontinuity in velocity. This discovery did much more than confirm the hypothetical model of the gravimetrists. It validated the idea of two distinct media separated by an *abrupt transition* (whereas measurements of gravity and of the moment of inertia could be interpreted as a slow, gradual, and continuous variation of internal properties, from an extremely light skin to an extremely heavy center). Moreover, Oldham's (1906) estimate of the core's radius was, as we shall see, very precise.

The work of Milne and Oldham immediately drew an enthusiastic response from scientists interested in the Earth. The method of measuring and analyzing transit times is easy to use and very powerful. Arrival times are plotted on a chart as a function of the distance from the focus of the earthquake (see Figure 5). The slope of the curve is directly proportional to wave velocity. In fact, this method was used for more than fifty years and was the basis of all the important discoveries about the internal structure of the Earth. The enthusiasm of the contemporary scientific community was reflected in the sudden multiplication of seismological observatories around the world and

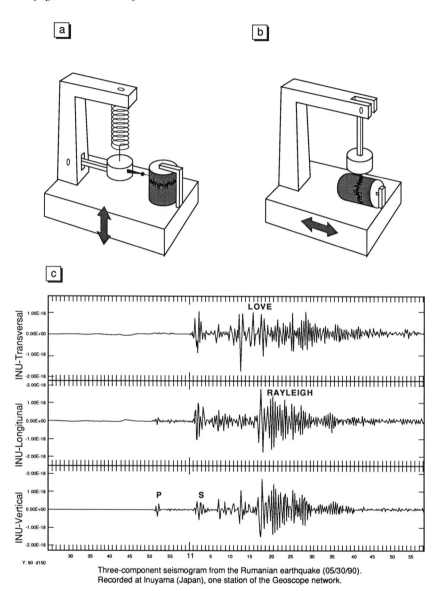

Three-component seismogram from the Rumanian earthquake (05/30/90).
Recorded at Inuyama (Japan), one station of the Geoscope network.

Figure 4 a) Seismograph recording vertical vibrations. b) Seismograph recording horizontal vibrations. c) Typical recording of seismogram in three seismographs representing one vertical and two horizontal vibrations.

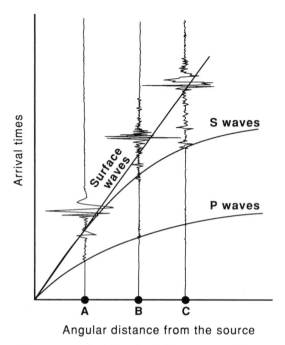

Figure 5 Travel time curve showing the different types of waves. Three seismograms from three stations A, B, and C are represented. Note that with distance the waves are more and more separated by the difference in their velocities.

in the increasing number of scientists working in the new discipline. All of them used this method to try to find anomalies or discontinuities in wave velocities and to discern internal structures. Results were not long in coming.

In 1909 Andreiji Mohorovičić, working at the observatory in Zagreb, Yugoslavia, showed that the outside layer of the Earth, the crust, is separated from the mantle by an abrupt discontinuity in the velocity of seismic waves. Soon the small seismologic community around the world confirmed this observation and gave it a general application. Since then, the discontinuity that separates the crust from the mantle has been called the *Mohorovičić discontinuity,* or more familiarly, the Moho. Under the continents it is usually located at a depth of 30 to 40 kilometers. In 1914 Beno Gutenberg, who was then working in Germany, refined the determination of the core-mantle boundary made by Oldham and calculated the radius of the core as 0.545 of the Earth's radius (at a depth of 2,900 kilometers), a value that is still accepted today (see Gutenberg, 1959; Bolt, 1982).

Thus, the basic internal structure of the Earth was well established before the First World War. This extraordinary development had taken hardly more than ten years. It had necessitated neither complex mathematics nor complicated theories but had applied the experimental method to the Earth using a simple technique of analysis and elementary geometric calculations. No doubt this episode would be better known and more widely appreciated by the scientific community in general if it had not coincided with the explosion of atomic and nuclear physics following Antoine-Henri Becquerel's discovery of radioactivity and Max Planck's introduction of quantum theory, events that completely obscured the birth of seismology.

Spherical Symmetry and the Egg Model

Seismology's second phase was devoted to explaining the principal characteristics of seismograms observed by applying the physical theory of the propagation of acoustic waves in a solid medium. Although the results obtained were the product of many researchers, two figures dominated this period: the English applied mathematician Harold Jeffreys, a virtuoso in manipulating the equations of classical physics and a determined adversary of continental drift who took the aloof attitude of the theoretician, and Beno Gutenberg, a German who had emigrated to the United States, where he founded the Seismological Laboratory of the California Institute of Technology (Caltech) with Charles F. Richter. Gutenberg's talent expressed itself in his detailed reading of seismograms and in his simple calculations, in which his physical intuition complemented mathematical developments. The two men disliked each other and spoke only during scientific meetings in spite of the efforts of common friends who tried in vain to get them to communicate outside of their famous public confrontations. Both identified the waves that, after being propagated through the Earth, arrive at a station and are registered on a seismogram. Their work made it possible to refine the image of the Earth's interior that had been developed by Milne and Oldham. They showed that the *P* waves are *compression waves,* while the *S* waves are *shear waves* whose essential characteristic is their inability to pass

The Caltech seismologists. Caltech's seismological laboratory has been the center of modern seismology for about forty years. Founded by Beno Gutenberg, it is now directed by Hiro Kanamori. From left to right: Frank Press, Gutenberg, Charles Richter (the inventor of the Richter scale), Hugo Benioff (discoverer of the Benioff zone).

through liquids (the absence of the *S* wave on a seismogram reflects the existence of a pocket of liquid on its path); that the paths of these waves are multitudinous and complex and that numerous variants for a given earthquake are found upon arrival at the station; that the surface waves are also complex; and above all, that a seismogram is a complex combination of all the waves that are superimposed upon each other, cancel each other out, and reinforce each other, making the reading of the recordings a real art, or, more exactly, the deciphering of a coded language. These theoreticians of seismology established the essential elements for decoding these complex messages.

The long effort in fundamental seismology gradually opened up our knowledge of the Earth's interior through a series of important

discoveries. The most fundamental accomplishment of the application of Milne and Oldham's approach was that it established a relationship between seismic velocity and the physical properties of the
medium. Thus, the fact that *S* waves cannot propagate through a
liquid allowed Inge Lehman (1936), who was then working at the
Copenhagen observatory, to assert that the outer core is liquid, while
the inner core is solid, a discovery Jeffreys and Gutenberg rapidly
confirmed. At the same time, the fact that outside of the restricted
area of the outer core *S* waves are propagated freely put an end to
the myth of the liquid mantle—an incandescent magma ready to
surge up under our feet at any moment. The vast majority of the
Earth's interior is indeed in the solid state, so a piece of the mantle
must melt before it can give birth to a volcano.

Using their curves of arrival times, Milne and Oldham had divided
the Earth's interior into a dense core surrounded by a less dense mantle. The study of surface waves allowed theoretical seismology to
refine this picture and draw a profile of density and seismic velocity
for the Earth's interior (see Figure 6). This picture showed slow, con-

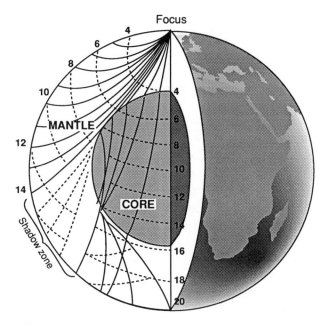

Figure 6 Different paths of seismic rays in the interior of the Earth. Notice
that the rays are curved and can be reflected by discontinuities in structure.

tinuous, and progressive progress and abrupt discontinuities in certain zones: at the base of the crust (the famous Moho), at about 400 kilometers, at about 650 kilometers, and especially at the core-mantle boundary, where the speed of the *P* waves changes abruptly from 9 kilometers per second to 13.2 kilometers per second or 47,000 kilometers per hour, the equivalent of going once around the Earth in an hour.

These discontinuities in seismic velocity correspond to discontinuities in density. If an increase in pressure causes atoms to move closer together and thus increases density, the general tendency—a slow increase in density as one moves deeper into the interior—seems reasonable (see Figure 7). Discontinuities—zones where seismic velocity and density increase abruptly—pose more complex problems. Does each discontinuity represent a sudden change in chemical composition? If so, then the Earth must consist of a series of layers differing in chemical composition. Or is it simply a question of a change in physical properties with chemical composition remaining constant (a *phase change*)? In that case, the globe would be chemically homogeneous, but the physics of matter under high pressure would have to be carefully explored, with the promise of interesting discoveries, since a *continuous* increase in pressure and temperature would correspond to *discontinuous* variations in the physical properties of matter. Geophysicists were to debate this puzzle for the next fifty years.

In order to picture the structure of the Earth as seismology reveals it, we can say that it consists of a series of nested spherical layers. At the center sits the *dense core,* with a high propagation rate and a solid inner nucleus. The *outer core,* with a radius of 3,500 kilometers, is liquid. Surrounding the core, with a thickness of 2,900 kilometers, is the solid but much less dense *mantle,* whose wave propagation velocity is much smaller than that of the core. Near the surface is the *crust,* a thin solid skin made of relatively light materials. Within the crust there are two clearly distinct areas: the oceans, whose crust is 5 kilometers thick, and the continents, whose crust varies in thickness because of mountains but whose average depth is about 35 kilometers. The structure of the Earth is thus like that of a hard-boiled egg, the crust corresponding to the shell, the white to the mantle, and the yellow to the core.

However precisely seismology has allowed us to determine the

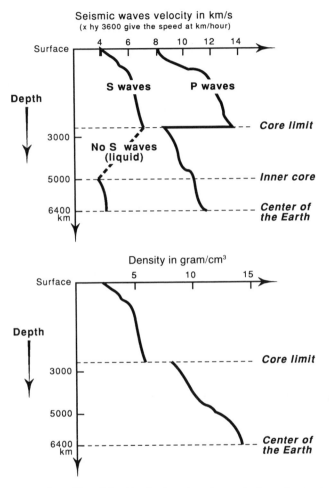

Figure 7 Internal profile of the Earth showing the variation of seismic velocity and density with depth. Notice the large discontinuity at the core-mantle boundary, the absence of S waves in the outer core (it is liquid), and the progressive increase of density with depth.

internal structure of the globe, we must admit that techniques based on gravity measurements and the moment of inertia had already given us quite a precise picture. If we look at the representation of the interior that was accepted at the end of the nineteenth century, before the development of seismology, and if we grant that scientists' confidence in this model and in the volumes assigned to the various layers has increased, we must concede that the earlier gravimetric model has not been fundamentally challenged.

The confirmation of the gravimetric model by seismology should encourage us to speculate on the internal structures of the other planets later on, although, in most cases, we know only their *average density* and their *moment of inertia*.

Translating the Message of the Waves into Mineralogical Language

Mapping seismic velocity or the densities of the Earth's interior is certainly an important step in understanding the Earth's structure, but it does not fully satisfy geologists. They want to know if the interior consists of granite, basalt, iron, or compressed helium. They want to understand it in terms of materials and chemical composition. The geochemist also wants to know the precise chemical composition of the Earth. So a translation, a code for transforming the seismological indices into material terms, had to be found. As we will see, this process was not simple or unique. A variety of approaches, numerous cross checks, and frequent controversies marked the development of the rather complete and coherent picture of the Earth's interior that we have today.

First, it took time to understand that solids subjected to the colossal pressures that prevail at the center of the Earth *are* compressible: their volume can be decreased under very high pressure and they can therefore increase in density. The common sense of elementary physics tells us that only gases are compressible, and it is this error of reasoning that made Benjamin Franklin believe the core was gaseous. After de Lame and de Cauchy's calculations at the beginning of the nineteenth century we had to wait for the high-pressure physics experiments to which Percy Bridgman's name is still linked to fully explain this phenomenon. Obtaining pressures of several kilobars, then of dozens of kilobars, and eventually of hundreds of kilobars (and today of megabars) is an extremely difficult operation in the world of the laboratory: materials break, gaskets give out, apparatus becomes unreliable; and when you actually reach these high pressures, you may not know how to measure them or what the temperature is. Thanks to a few pioneers, however, this new branch of physics began to develop.

Seismologists such as Gutenberg rapidly understood how interesting high-pressure physics was for seismology. Interest remained at a
purely formal level, however, until a person capable of bridging the
two disciplines appeared. He was Francis Birch, a pupil of Bridgman
at Harvard, where he would spend his entire career. In close contact
with the seismologists, he attempted to measure seismic velocity in
various materials and under a variety of conditions of temperature
and pressure. He established the relationships between seismic velocity and density, and between pressure, temperature, seismic velocity,
and chemical composition. This was a systematic, meticulous, thank-
less task that brilliantly illuminated the seismic data. These data,
illustrated in Figure 8, show that the mantle is undoubtedly rich in
silicon, while the core must be rich in iron because under high pressure the density of iron rises very quickly, reaching 11 to 13 grams
per cubic centimeter. The presence of rare metals such as gold or
platinum is therefore not necessary to explain the density of the core:
iron, a common metal, suffices. *The identification of the core as compressed iron is incontestably one of the major discoveries of geophysics.* Its

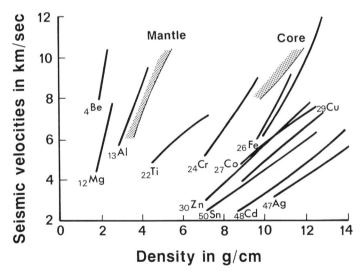

Figure 8 This figure illustrates Francis Birch's lab experiments. Each curve
is the variation of seismic velocity with density. The hatched areas are the
values measured by seismologists. Birch's interpretation of this figure can
be simplified as mantle equals silicates, core equals iron-nickel alloys.

significance and consequences will become clear over the course of this book (Birch, 1961). Further studies showed that to obtain a precise agreement between seismological measurements and laboratory experiments it must be assumed that the core is composed of an alloy of iron and nickel—the famous NiFe dear to prewar geophysicists.

The exact nature of the terrestrial crust was not much debated: surface rocks are in fact observable, and tectonic accidents and deep mountain valleys expose pieces of crust to a depth of several kilometers. It soon became clear that there was a fundamental difference between oceanic crust and continental crust. Oceanic crust, whose thickness is 5 kilometers, consists of basalt, that is, volcanic rock. The upper continental crust consists of rock rich in silica, that is, granite, while the deeper zones of the continental crust must consist of a mixture of basalt and granite. The terrestrial crust as a whole is thus composed of lighter materials than is the mantle, the continental crust being lighter than the oceanic crust.

Because the mantle is deeper and therefore farther away, its nature is more difficult to determine than that of the crust. Birch's experiments showed that the mantle, like the crust, consisted of silicate materials. If the mantle is a rocky medium, the question is: what kind of rock? After quite a technical debate, this problem was solved by combining information from seismic measurements and laboratory experiments. Today no one doubts that the vast majority of the upper mantle consists of a rock called *peridotite.* The crust-mantle boundary, the Moho, corresponds to a change in the chemical composition of the rock. We would do well to remember the name *peridotite,* since we will encounter it often and it deserves to be as well known as basalt or granite.

The nature of the deep mantle and the discontinuities of seismic velocity that had been observed at 450 and 650 kilometers were more difficult to establish. Today it is thought that the entire mantle has a chemical composition—in particular a richness in silicon and magnesium—similar to that of the upper mantle, although its mineralogical composition is very different. The minerals that constitute it must be denser than those in the crust as a result of the increase in pressure. So seismic discontinuities must occur at the depth (and therefore the pressure) at which the mineralogical changes take place.

But this is an area of active research that should not be cut off by

premature conclusions. In any case, the development of new laboratory techniques for very high pressures should allow us to solve
these problems soon. By squeezing two little diamond blocks in an
anvil cell it is possible to reach pressures of several megabars, which
is several million times atmospheric pressure. The compressed object
can be heated by a laser ray that passes through the diamonds, making it possible to reach a temperature of thousands of degrees and a
pressure of several megabars at the same time—experimental conditions similar to those in the Earth's interior. This is technical prowess
unimaginable ten years ago: the conditions of pressure and temperature at the Earth's center can now be reproduced in the laboratory.

To recapitulate the results of this dialogue between seismological
observations and laboratory experiments, the chemical structure of
the Earth could be described as follows (Ringwood, 1975): At the
center there is a core of nickel-iron with a density of 13 grams per
cubic centimeter. This is surrounded by a silicate mantle of rather
homogeneous chemical composition 2,900 kilometers thick with an
average content of 20 percent silicon and 24 percent oxygen and a

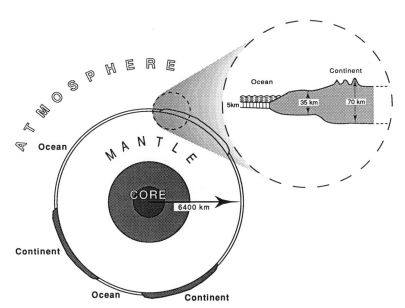

Figure 9 A cross section of the Earth showing the egg model and its
envelopes (a section of the crust is enlarged on the right).

density of 4.5 to 3.2 grams per cubic centimeter, which contains magnesium, iron, and a little calcium and aluminum. On top of all this is the crust, which can be divided into two provinces: the basaltic oceanic crust consisting of 50 percent silicon dioxide with a density of silicates of calcium and magnesium of 3 grams per cubic centimeter, and the continental crust, which is lighter and richer in silicon (more than 30 percent). In addition to silicon the continental crust contains mainly aluminum, potassium, and sodium.

The solid Earth is itself like an egg in another sense: it is surrounded by a gaseous envelope, the atmosphere, and covered by a film of water, the oceans (see Figure 9). The atmosphere consists of nitrogen and oxygen, the oceans of hydrogen and oxygen. The principle that density decreases toward the exterior applies not only to the internal layers but also to the external envelopes.

The Envelope Structure and the Formation of the Earth

The discovery that the Earth consists of successive envelopes of differing chemical composition raises the question of how our planet formed and evolved in a striking way. If the Earth were of uniform and homogeneous chemical composition, the only question would be how these materials accumulated to form a great ball. But the envelope structure—the presence of a heavy core surrounded by a lighter mantle and crust, the existence of a gaseous atmosphere and a superficial layer of water—demands an explanation of how and why this chemical differentiation into layers took place. In what period were the atmosphere and the oceans formed? When did the core and the continents differentiate themselves? Under what influences did such differentiation take place? What energy sources were used? Every structure invites inquiry as to its genesis, Earth more than any other.

As soon as the internal structure of the Earth was known with an acceptable degree of probability, geophysicists began to wonder about its formation and evolution. This questioning was even more natural for researchers who had been trained as astronomers, such as the famous Harold Jeffreys. Little by little these questions led to the development of two opposed hypotheses: *heterogeneous accretion* and *homogeneous accretion*.

The Heterogeneous Accretion Scenario

According to the heterogeneous accretion scenario developed by Karl Turekian and Sydney Clark of Yale, the solid materials that constitute the Earth—dust or rocks—assembled or "accreted" in the order of their density. Those that were heaviest, such as iron, assembled first, creating the core. Slightly lighter but still solid materials such as the silicates accreted around them to form the mantle and the crust. Finally, gaseous materials such as air and water were captured by this large rocky mass and formed the oceans and the atmosphere (see Figure 10). In this model the Earth's layered structure is as old as the Earth itself. It is not only intimately related to the manner of its for-

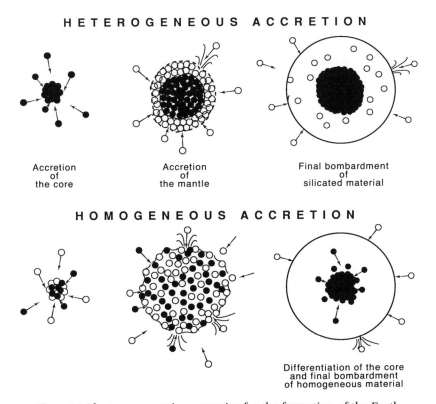

HETEROGENEOUS ACCRETION

Accretion
of
the core

Accretion
of
the mantle

Final bombardment
of
silicated material

HOMOGENEOUS ACCRETION

Differentiation of the core
and final bombardment
of homogeneous material

Figure 10 The two competing scenarios for the formation of the Earth.

mation but results directly from it. Geology can consider the global structure of the Earth a basic given.

The Homogeneous Accretion Scenario

According to the homogeneous accretion scenario, the accretion of the terrestrial materials began with a homogeneous cloud of dust. The Earth of the earliest days was a ball that everywhere, in the center as well as on the edges, contained the same proportion of iron as of silicates and of silicates as of water. Following its accretion the Earth differentiated itself—differentiation is a fundamental geological concept—and engendered the various envelopes that we know today. No doubt it passed through a hot, molten stage during which the heavier iron fell toward the center, while the silicates remained near the edges. Volatile products escaped towards the surface, forming the atmosphere and the oceans.

Metallurgical processes support this theory to a striking degree. When minerals are melted in the presence of carbon to create a reducing atmosphere, the heavy metallic iron sinks to the bottom, the silicated mass floats, and the process is accompanied by an abundant gaseous discharge. As the German geochemist Victor Goldschmidt noted in 1940, this is a small-scale picture of what may have happened when our planet was formed.

In the scenario of heterogeneous accretion, accretion and differentiation take place simultaneously. In homogeneous accretion they succeed one another: an initial hot phase is followed by the progressive establishment of present conditions through cooling. Heterogeneous accretion created a world almost identical to the one we live in, and Lyell's principle of uniformitarianism can be applied to it without too much difficulty.

The origins of the atmosphere and the ocean are totally different in the two scenarios. Proponents of homogeneous accretion believe that the atmosphere and the ocean are the result of degassing from the mantle; their formation was an integral part of the primary process of differentiation, as was the formation of the core, and in a somewhat symmetrical way. In contrast, according to heterogeneous accretion the ocean and the atmosphere were captured by the Earth either in the form of previously heated ice or as dense clouds. They were never in contact with the mantle or the core.

How to choose between these two models with their fundamentally different consequences? Investigated at length according to the principles of physics, both are equally plausible; neither violates its laws. Astronomy offers almost no information. Therefore we must turn to other methods to test these scenarios. To understand the resolution of this controversy we are obliged to look beyond the Earth and toward other planetary objects, and to examine the Earth with greater attention and with more penetrating methods.

But a simple yes or no answer cannot be expected. Einstein used to say that when you question Nature she usually answers perhaps. We could add that not only does she not clearly answer the question asked, she usually reveals new questions that are equally fascinating and puzzling. This is the very essence of the scientific process. In the previous chapter we saw that the study of surface geology led to the idea of "no vestige of a beginning"; further study of the gross structure of the Earth using geological methods provides evidence of an early episode in the formation of the planet. We are then left with questions: What was that episode and how did it come about?

3

The Geologic Calendar

There can be no serious history without serious chronology. Since its emergence, geology has attempted to establish a precise geologic chronology, but it was a long time before those in the field were able to develop a way to measure geologic time using numbers. We will follow the historical progress that led, little by little, to the establishment of a geologic calendar, one of modern science's most important frames of reference.

The first geologic chronology was taken directly from the Bible. According to the book of Genesis, the world was created in six days, Man last of all. Since the history of Man was measured in thousands of years, Earth's history must also be measured in thousands of years. Around 1650 James Ussher, an Irish prelate and scholar, established a "geological"chronology according to which the Earth had been created on October 26 of the year 4004 B.C. at nine o'clock in the morning (Faul, 1978; Hallam, 1983). At the time, this calendar was considered very accurate because it was based on a close reading of Greek, Egyptian, and Christian manuscripts. The creation of the Earth, which was naturally confused with the creation of the universe, had been followed by the creation of Man.

Werner's Neptunian theory and his six stages of geologic development must be placed in this chronologic context. We know that Hutton, Playfair, and then Lyell, however, utterly opposed the idea of geologic time as short and *finite*. Their theory of Uniformitarianism gave geologic time an *infinite* duration. Just as the motion of the planets seems to have neither a beginning nor an end, geologic phenomena repeat themselves over an infinitely long period of time. These phenomena are permanent features of nature. Infinite time is an extremely useful hypothesis in geology because it completely eliminates the idea of an exact age of the Earth and therefore the need for the absolute measurement of geologic time. Since phenomena repeat themselves over and over, it is neither necessary nor important to establish a chronology which, in any case, would be lost in infinity. Although Hutton was correct in his analysis of geologic causality, his views long warped our understanding of time.

We know that Georges Buffon (1749–1783) had already grasped the problem as a whole. While defending the theory of uniformity of process like Hutton, he had also understood the necessity of being an evolutionist like Werner; he estimated that the Earth was 200,000 years old. So the focus of the debate became infinite, cyclic, repetitive time versus finite, directional, and measurable time. This is a geologic transposition of the legendary debate between the cyclical time of the Egyptians or the Taoists and the vectorial time attributed to Judeo-Christian tradition, which has a beginning.

To move forward on such paths was to risk provoking heated religious debate and anger. At the same time, from a purely technical point of view geologic science could not refuse to take up the problem of estimating geologic time precisely because without precise time, the whole question of the sequence and duration of geologic events would stagnate in a morass of relative and qualitative chronology.

Stratigraphy and the Geologic Time Scale

As its name indicates, stratigraphy is the study of *strata* or geologic beds (Toulmin and Goodfield, 1965; Hallam, 1983). Deposited one on top of another at the bottom of the sea, the strata constitute a multilayered book, each page of which represents an episode in the Earth's history. Since the Renaissance and thanks to the efforts of men like Leonardo da Vinci and Nicolaus Steno, it has been understood that the strata had been deposited on the sea bottom and that the order of their deposition must be read from lowest to highest. These scientists grasped the idea, although in a hazy way, that a given thickness of terrain corresponded to a time interval, the *deposition time.* But in those days there was no way to compare a series of strata from one place with a series from another place, one from the Paris basin for example, with one from the London basin.

The significance of the fossil had to be understood before further progress in the exploration of geologic time was possible. The geologic use of fossils began in the second half of the eighteenth century. It appears that Antoine Lavoisier (see Dott and Batten, 1981), who would become famous as a chemist, was the first to use the principle of correlating sedimentary strata by fossils. In 1789, five years before his execution during the Reign of Terror, Lavoisier published a series of articles in which he showed in clear diagrams that each series of geologic beds contains a set of characteristic fossils and that geologic strata can be identified regardless of their geographic position; that is, two strata hundreds of kilometers apart can be identified as deposits laid down at the same time period (see Figure 11). It appears, however, that his work went unnoticed, and the paternity of the science of stratigraphy is generally attributed to Georges Cuvier and Alexandre Brongniart (1808) on the one hand, and to William Smith (1817) on the other.

The principles are rather simple. Because fossil fauna change over the course of time, fossil succession can be established by studying fossil variations over time in a given place. Geologic formations can be correlated by studying the strata in different places. By establishing strata successions that contain fossils in one locality, and then translating such scales to other places and repeating the process several times, one can gradually built a stratigraphic scale (see Figure 12). Using this principle Smith drew up a remarkable series of geologic maps for central England, as did Cuvier for the Paris basin. Cuvier then tried to interpret his observations of the changes in the fauna and proposed, as we have seen, the theory of catastrophes, the source of a fierce dispute with the Uniformitarians.

This dispute among "theoreticians" did not, however, prevent stratigraphy from continuing to develop. In England, Sedgwick, Conybeare, and Murchison established the stratigraphy of terrains they began to call primary. These were the sedimentary terrains lying directly upon the crystallized terrains in Scotland. French and German geologists developed the stratigraphy of secondary and tertiary terrains, that is, those strata lying above primary terrains. After 1860 it became possible to identify and correlate sedimentary terrains all

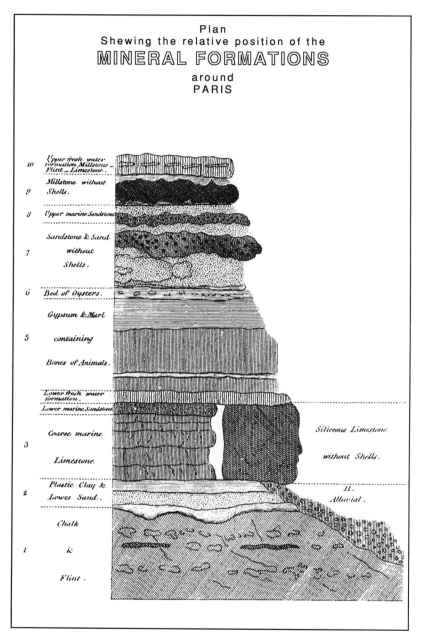

Figure 11 The first stratigraphic section, drawn by Antoine Lavoisier.

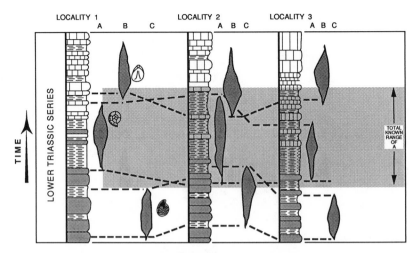

Figure 12 The principle of stratigraphic correlation based on fossils. The time of the appearance or disappearance of each fossil is shown by the dotted line. Notice the uncertainty of correlation (time increases going upward).

over Europe. After 1880, as a result of the efforts of Charles Walcott especially, American terrains could also enter what was beginning to be called the *scale of geologic terrain.* It is now accepted that this scale, each stage of which corresponds to a certain thickness of sedimentary rock, in fact reflects a temporal division. The equation *thickness of terrain = time* gradually led geologists to substitute the term *geologic era* for the word *terrain:* they spoke of primary terrain or the primary era, of secondary terrain or the secondary era. They subdivided the eras into periods. The scale of terrains soon came to be called the *scale of geologic time,* a significant semantic shift if ever there was one. The scale remained purely relative, however, and no one dared to guess the duration of the various periods, nor did anyone dare to say whether the scale had a lower boundary or whether it was simply a little slice of infinite geological time, the slice Nature had graciously left for us!

Lord Kelvin and the Short Chronology

At the end of the nineteenth century, however, the idea of calibrating the geologic scale with units of time was discussed in scientific circles, showing that scientific thinking had been evolving, even among those who propounded Hutton's theories. It seemed that geologic time could be divided into two great periods: *ancient* time, in which rocks without fossils were formed (because these lie under the Cambrian strata and are therefore anterior to the Cambrian, they are called Precambrian), and *geologic* time, whose fossil-containing terrains are further divided into the Paleozoic, Mesozoic, Cenozoic, and Quaternary. This second period is the geologically "interesting one," because it is the only one for which a rigorous geology based on both geometry and paleontology can be practiced.

Sensing the question in the air, various geologists nevertheless tried to "calculate" the durations of the geologic periods and to calibrate the time scale. In 1859 Charles Darwin, who was a geologist as well as a biologist, made a quick (and false) calculation and estimated that it had taken 300 million years to carve out the Wealden Valley in the southeast of England. The scale of "geologic" time was therefore long. In Ireland, John Joly calculated the age of the ocean using an ingenious method. According to widespread belief, the salinity of the ocean was due to evaporation, which concentrated the salts supplied by rivers and streams (Toulmin and Goodfield, 1965). (In fact, the ocean's salinity is determined by a complex phenomenon to which we will return later.) Calculating the flux of salt added annually and the salt already contained in the oceans, Joly concluded that it had taken at least 100 million years to reach the present salinity. The Earth must therefore be 100 million years old, including the Precambrian period, and the time scale must be rather long.

Clearly these methods of calculation are strict applications of Lyell's principle of uniformity of process, since they extrapolate from the duration of present-day events. Scientists continued to ignore the concept of infinite time. We might note that the figures used are all greater than 10 million years. *Million* as a unit of time entered the geologic vocabulary at this time. The majority of geologists, following Lyell's intuition, concluded that all these figures were at most useful *minima* for estimating the length of a geologic cycle and that the actual duration of geologic time was much longer. They spoke of several hundreds, thousands, even millions of years. For them geologic time was perhaps not infinite, but it was certainly very long.

To counter this view of almost infinite geologic time in the name of the fundamental principles of physics and of the law of conservation of energy, Lord Kelvin initiated a struggle in 1846 that lasted more than fifty years. The energy of a system is finite; if its energy dissipates, its "activity" must also decrease over time. The Earth, Kelvin said, is losing heat. In fact, when one descends into the terrestrial depths (for example, into a mine), the temperature increases. According to Fourier's laws of heat propagation, heat propagates from warm to cold and therefore from the Earth's interior to its surface. If the Earth is continuously losing heat, its internal activity must decrease over time. The idea of cyclic geologic activity in an Earth of infinite age is absurd, Kelvin said (Kelvin, 1899; Burchfield, 1975; see also Hallam, 1983).

Kelvin decided to calculate the age of the Earth as a quantitative expression of his deductions. He estimated the time it took to cool the terrestrial surface from 2,000°C or 1,000°C to today's 25°C, taking account of the heat flow that the Earth generates. The resulting figure was 100 million years, which agreed with the result of a similar calculation he had made for the sun. He therefore concluded that the Earth and the sun were formed at the same time, about 100 million years ago. Returning to his calculations several years later and realizing the uncertainties that his hypotheses implied, he widened his margin of error and concluded that the Earth was between 25 and 400 million years old.

Around 1880 Krieg and Carl Barus's first precise measurements of the thermal conductivity of rocks led them to reevaluate the heat flow and to lean toward a short chronology. They estimated the age of the Earth as 25 million years, and Kelvin went over to their position.

Aware that Kelvin's criticisms and calculations could upset his theory of uniformity, Lyell attempted to respond. He invoked the creation of energy in the Earth's interior (and also the sun's) through chemical reactions. Impervious to this brilliant intuition, Kelvin answered that that was exactly the sort of argument advanced to justify research on perpetual motion.

Since Lord Kelvin was the greatest physicist of his time, no physical scientist dared to criticize his reasoning or his figures. Geologists alone continued to believe in long durations, but on such subjective grounds that they could not convince the rigorous scientific minds of their day. The attentive scientific spectator was forced to choose between a scientific argument and a conviction, a rigorous demonstration and a qualitative impression: between the proud science of physics, full of the successes of Newton, Maxwell, Fourier, and Carnot, and geology, a discipline that writers since Buffon's time had located at the junction of science, literature, and philosophy.

As often happens in science, light finally dawned in the distance. The debate between long and short chronology started by Werner and Hutton at the end of the eighteenth century had still not been resolved a century later. The need for quantification, however, had been made clear to everyone. The choice was no longer between a million years and infinity but between millions of years and billions of years.

The Revolution in Radioactivity

In 1896, in a shed at the far end of the Jardin des Plantes in Paris, Becquerel discovered that uranium salt emits strange rays, which darken a photographic plate. Within a few years a group of pioneers, including Pierre and Marie Curie in France, Ernest Rutherford and Frederick Soddy in the United Kingdom, and others, discovered the key to this mysterious phenomenon.

Some atoms decompose, disintegrating spontaneously into other atoms. In other words, alchemy's old dream of transforming one element into another occurs spontaneously in nature. But it does not take place on demand. Only certain atoms, such as uranium and radium, have this property and, as the pioneers soon discovered, it seems to be intrinsic to the atom. It takes place invariably, no matter what physical conditions surround the atom or whether it is free or locked in a chemical combination.

This is the domain of fundamental physics and seems a long way from geologic concerns. But we must remember that the scientific mind at the beginning of the twentieth century was still encyclopedic in scope, or at least very curious about problems in other sciences. In fact, the transfer of information between the new physics and traditional geology was to take place with surprising speed. Lord Kelvin's conclusions were called into question again by two totally different discoveries.

As soon as Pierre Curie and Laborde discovered in 1903 that radioactive disintegration creates heat, Rutherford (1906) began to envision the consequences of this discovery for the thermal history of the Earth. The Earth's interior, like its surface, contains radioactive elements, especially uranium. The basic assumption of Lord Kelvin's calculations—that the Earth, which was initially hot, is cooling inexorably—must be questioned again, since the Earth possesses its own heat source (Lyell's intuition was right). The age of the Earth calculated by Lord Kelvin must therefore be only a minimum, without any real significance.

At first, by invalidating Lord Kelvin's estimates, the discovery of radioactivity destroyed the quantitative method of calculating time. Later it led to a substitute method that was more powerful and more rigorous because it made actually measuring time possible.

Rutherford, a young physicist from New Zealand who had emigrated to Canada and then to England, played a central role in all these developments. He was the first to propose the daring hypothesis that the proportion of radioactive atoms that disintegrate in a given time interval is an unvarying constant and therefore a potential clock (Rutherford, 1906). If, for example, you have 10 billion radium atoms, 1,622 years later only half of them will be left, and in 3,244 years, only a quarter. Thus, the mass of radium is reduced by half every 1,622 years (see Figure 13). The disintegration of radium follows an exponential law. Since the quantity of radium, uranium, or any other radioactive element decreases in a simple way, by measuring that decrease it is possible to measure the passage of time. Since

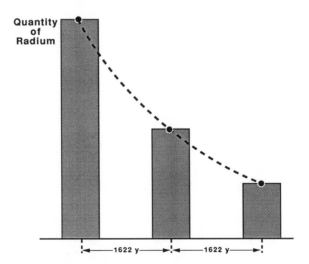

Figure 13 The principle of radioactivity. During each period the quantity of radium has decayed by half.

the decay rate of certain elements such as uranium is measured in hundreds of millions of years, this phenomenon can serve as a clock for measuring geologic time.

There is, however, a problem to be solved before this principle can be used. It is certainly possible to take a mineral or a rock and measure the quantity of uranium it contains. But how do you determine the quantity of uranium that mineral or rock contained when it was formed (an indispensable fact if you want to subtract to find the amount that disintegrated)? Rutherford proposed the *residual* method to solve this problem (see Figure 14). When you want to

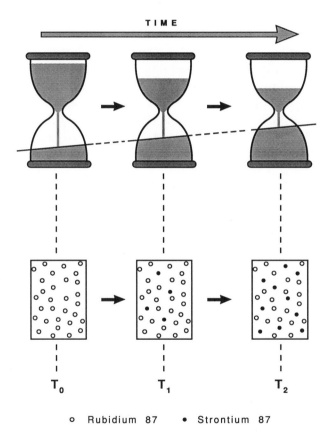

○ Rubidium 87   ● Strontium 87

Figure 14 The principle of the radioactive clock. With time, the upper part decays while the lower part increases as rubidium transforms into strontium-87; the number of rubidium atoms decays with time, while the number of strontium-87 atoms increases. Measuring the ratio of rubidium-87 to strontium-87 provides a way of calculating age.

know how long it will be before an hourglass runs out, it is not enough to look at its upper compartment, because you don't know how much sand it previously contained. But if you look at the amount of sand contained in both the upper part and the lower part, you can deduce how much time has passed since the hourglass was turned over. Rutherford applied the same principle to the decay of uranium. When uranium decays, it produces alpha rays, otherwise called helium atoms, a gas that William Ramsay had just discovered in the atmosphere. Since uranium (actually uranium and its radioactive daughters) produces eight helium atoms each time it decays, if you know the number of uranium atoms that have decayed, you can calculate the number of helium atoms. Since you can measure the amount of uranium that is still present, you have access to the proportion that has disappeared, and therefore to a time measurement. By simultaneously measuring the quantities of uranium and helium present in a mineral you can apply the hourglass method and calculate the age of the mineral.

Rutherford applied his new method to a series of uranium minerals whose helium had been measured by Ramsay. The resulting figure was 1 billion years, then 1.5 billion years. For the first time the age of a mineral was measured *directly,* and this age, which must be a minimum for the age of the Earth, proved Lord Kelvin wrong. A billion years won out over a million years!

At about this time the chemist Bertram Borden Boltwood (1907) was working at Yale University. Becoming interested in the "radioactive alchemy" that allows one element to transform itself into another, he made a systematic chemical analysis of uranium ores to discover whether they contained certain chemical elements in unusual abundance. He noticed that lead seemed abnormally rich in the uranium ores and hypothesized that lead is the ultimate product of the decay chain of the uranium family, certain elements of which had been discovered by Pierre and Marie Curie. Boltwood, who was extremely interested in Rutherford's method, told him about his idea, and Rutherford suggested testing his hypothesis by a method that could be called geologic. If the radiogenic lead (that is, lead "generated" by radioactivity) hypothesis is correct, lead, like helium, must be accumulating. The amount of lead must therefore be greater in old uranium ores than in young ores. Cooperating with the geology department at Yale Boltwood rapidly verified Rutherford's suggestion and using the principle in reverse determined the ages of numerous uranium ores. Depending on the ore, he found ages of 200 million to 2 billion years; what is even more important, those ages were consistent with the geology. They were older when the ore was found in the basement cover of Phanerozoic sediments and younger when the mineral was found above the Phanerozoic sediments. These discoveries gave geology a quantitative historical context for the first time and immediately aroused geologists' interest. By 1917 so many age determinations had been made for lead and helium that a considerable collection of geologic ages existed.

At that point Joseph Barrell, a geology professor at Yale, undertook the first synthesis of all these absolute ages. First he did a critical analysis of all the published results, placing them in their geologic context and retaining only those that did not violate fundamental principles of geology (for example, the sediments that cover a basement should be younger than the basement rock itself). Using this tool he estimated the time durations that corresponded to 100 or 1,000 meters of sedimentary strata for various geologic periods. In other words, he calibrated in time the sedimentation process that generated the sedimentary strata. Then he attempted to erect an absolute scale for the geologic eras based on this calibration, the first of its type (Barrell, 1917; see also Hallam, 1983). Let us stop a moment to look at it.

Barrell placed the beginning of fossil time—the first terrains in which fossils are found, which are called Cambrian—between 550 and 200 million years ago (today we know that this period in fact began 550 million years ago), the end of the Paleozoic era at about 215 million years ago (today it is placed at 230 million years), and the length of the Quaternary period at 1.5 million years (today we would say 3 million years). If we compare this study with what has been patiently established since, we realize with no small astonishment that, although few geologists are aware of it, Barrell took the decisive steps in the establishment of the geologic calendar in 1917.

Although this scale held considerable potential for geology, it was not really accepted or used by geologists until 1955, almost forty years later; at the same time—and this is what concerns us here—it indicated nothing about the Earth's age. Barrell's scale of geologic time certainly ended the quarrel between long chronology and short chronology, but it contributed nothing to the debate over the concept of geologic time. Was it cyclical and infinite, or vectorial and of limited duration?

Arthur Holmes and the Age of the Earth

The first real attempt to determine the age of the Earth directly was made by Arthur Holmes, whose name is still linked with the flowering of the historical dimension in modern geology. Holmes's (1911) approach was twofold. Like Barrell, he established a scale of fossiliferous geologic time, but then he pushed the exercise further. Having estimated the length of time necessary to deposit a given thickness of strata, he extrapolated—an undertaking even more risky than Barrell's interpolations—to obtain an age for the oldest terrains. He corroborated his calculations by using ages that he or Boltwood had obtained for ancient terrains that did not contain fossils. He then concluded that the Earth was at least 1.4 billion years old and less than 3 billion years old.

This touches on a principal difficulty in determining the age of the planet. The age of a rock or a mineral is a well-defined concept: it represents the length of time that has passed since this rock or mineral crystallized and consolidated. All you have to do to determine its age is to measure the quantities of uranium and lead the rock or mineral contains. But the age of the Earth? Can we hope to measure the global quantity of uranium and lead in the Earth? Was the Earth formed rapidly enough to be able to attribute a specific age to it?

The only approach that at first seems possible is to define a lower limit for the Earth's age: if the oldest rock is 2 billion years old, the Earth is more than 2 billion years old. Arthur Holmes wanted to do better. He postulated that terrestrial lead is radiogenic in origin, that is, it is produced by the disintegration of uranium. So originally the Earth must have contained no lead. Estimating the abundance of lead and uranium in standard rocks of the Earth's crust, he set the age of the Earth at 3 billion years. Since the two approaches furnished comparable figures, Holmes (1927) concluded that 3 billion years was the correct one. The question remained at this point until the Second World War.

Lead's Memory

Common parlance doesn't attribute any of the elephant's characteristics to lead. One speaks of a leaden sun, of objects as heavy as lead, but never of a memory like lead. Yet this special quality made decisive progress in calculating the Earth's age possible. Since Boltwood's work it had been known that uranium gives birth to lead as the result of a series of disintegrations. The truth is actually a bit more complex. To begin with, there is not just one type of uranium, but two. They both have the same chemical properties, but the structure of their nucleii, and therefore their mass, is different. Their names, uranium-238 and uranium-235, are based on their mass. Both are radioactive, but they disintegrate at different rates. The decay rate of uranium-235 is twenty times faster than that of uranium-238. The situation at the other end of the radioactive decay chain is also complicated. There is not just one type of lead, but four, each with a different mass: lead-204, lead-206, lead-207, and lead-208. Patient research has made it possible to join the two ends of the radioactive decay chain correctly. Uranium-238 gives birth to lead-206 when it decays; uranium-235 to lead-207. Lead-208 is produced by the decay of another radioactive element that is similar to uranium, thorium. Lead-204 is the result of no known decay process.

Thus the situation is infinitely more complex—but also richer in possibilities—than Boltwood had envisioned. Naturally occurring uranium is a mixture of the two uraniums with different radioactive characteristics. Natural lead is a mixture of four different leads, two of which are related to uranium through radioactive decay. The relationship between uranium and lead therefore contains not one radioactive chronometer, as Boltwood thought, but two chronometers closely linked in nature.

Before examining all the consequences of this characteristic of uranium-lead systems, we should look at how such a complex system was decoded. These discoveries all resulted from the use of a new instrument, the *mass spectrometer,* invented by F. W. Aston (1919), an English physicist. The mass spectrometer is an instrument that weighs atoms; in some sense it is an atomic balance. It utilizes the effect of a magnetic field on a bundle of ionized atoms to separate atoms of different masses. Aston discovered not only that the various atoms could be separated from one another, but also that several varieties of the same element that were chemically identical but different in mass could be distinguished from each other. The mass spectrometer also determined the proportion of each variety for each element, for example, the proportions of lead-204, lead-206, lead-207, and lead-208 contained in a sample of lead. Different varieties of the

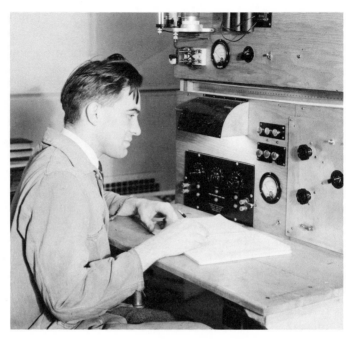

Alfred Nier at Harvard when, as a postdoctoral fellow, he
discovered the natural variations in lead isotopes.

same element are called *isotopes* of that element. Thus, uranium-235
and uranium-238 are two isotopes of the element uranium. Using
the same terminology we say that the mass spectrometer reveals the
isotopic composition of lead, or uranium, or any other element in
nature. Without the mass spectrometer, none of the decisive steps in
achieving our current knowledge of the universe would have been
possible.

Between 1936 and 1937 Alfred Nier, who was doing a postdoctoral
fellowship at Harvard University, applied these discoveries to refining
Boltwood's old chronological method. Using a mass spectrometer
and chemical analysis it is possible to determine how much of each
isotope of uranium and each isotope of lead is present in a given
sample of uranium. Using the residual method we can calculate two
geologic ages, one for the radioactive relationship between uranium-
238 and lead-206 and the other for uranium-235 and lead-207. Of
course, we can compare these two ages and use each to test the reli-
ability of the other, but even more can be done. If we have two hour-

glasses running out at two different known rates, determining the proportion of sand in the two lower reservoirs, even if we do not know how much sand was in the upper reservoirs, is enough to tell us how much time has passed. It is the same for lead. We can show that, in effect, determining the ratio of lead-206 to lead-207 allows us to measure time. There is no need to know how much uranium-238 and uranium-235 is present. What economy of technique! No more measuring the absolute quantity of uranium and lead: measuring the isotopic composition of lead on the mass spectrometer is all that is necessary (see Figures 15 and 16).

The internal isotopic analysis of lead is thus capable of revealing

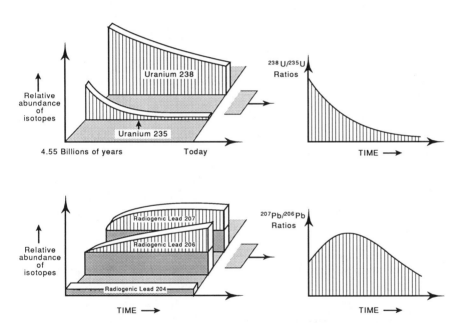

Figure 15 The uranium-lead radioactive system. Uranium-238 and uranium-235 are both radioactive, but uranium-235 decays much faster. Lead-206 is the final product of the uranium-238 decay chain, while lead-207 is the final product of the uranium-235 chain. Because uranium-238 undergoes uniform decay, lead-206 grows uniformly with time. In contrast, lead-207 increased quickly in the past and then much less so because uranium-235 is much less abundant. For this reason the isotopic ratio $^{207}Pb/^{206}Pb$ is a function of time and can be used as a chronometer. Lead-204 is used as a reference since it is not produced by radioactivity. The graphs show the variation in isotopic abundances of uranium and lead with time, assuming for the system an average $^{238}U/^{204}Pb$ ratio.

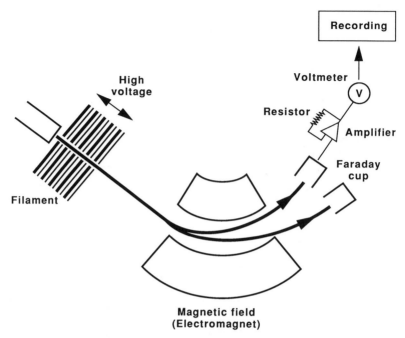

Figure 16 A schematic diagram of a mass spectrometer. Ions of one element are accelerated by high voltage in a vacuum. This stream passes through a magnetic field and is deflected. Depending on the mass of the ions the degree of deflection is more or less important (the heavier ions have the greatest inertia). Using this device we can separate the different isotopes of one element. A faraday cup measures the intensities of the different beams.

the age of the uranium ore that contains it. To demonstrate the validity of the method he had just invented, Nier measured the ages of a variety of uranium ores using the three chronological methods he now had at his disposal (the uranium-238–lead-206 method, the uranium-235–lead-207 method, and the new internal method for lead that would soon be called lead–lead). In most cases the ages obtained by the three different methods coincided. Radioactive chronology was therefore a reliable method, and the oldest age found—about 2 billion years—must be considered a minimum age for the Earth (Nier, 1938; 1939).

It would be convenient to be able to use the same method for the entire Earth, but how could the isotopic abundance of lead in the whole terrestrial crust be estimated? Nier considered analyzing a set of well-chosen rocks and averaging the results. Unfortunately, at the time the technical capability necessary to accomplish such a project did not exist. Lead is present in modern rocks only in very low concentrations, and it was not yet possible to analyze traces of lead in a mass spectrometer. Nier had to be content with analyzing galena, a lead ore that is found in terrains of various ages. He showed that the isotopic composition of lead in the ores varied with their age and the region in which they were found (Nier et al., 1941). He drew no conclusions, however, about the age of the Earth. War had just broken out.

The Age of the Planets

As the war was ending, various researchers, including Gerling in the USSR, Holmes in the United Kingdom, and Houtermans in Switzerland, attempted to use the isotopic lead method and the results obtained by Nier on lead ores to calculate the Earth's age. Their efforts, although they were important in many ways, hardly changed the situation: the estimated age remained around 3 billion years.

With the passage of time it has become clear that the most important breakthrough was that of Clair Patterson around 1950. Patterson's first advance was a technical one. With George Tilton at the University of Chicago, where they were both students at the time, he developed a technique for analyzing the isotopic composition of microquantities of lead. He could therefore determine the isotopic composition of the lead in rocks (more common than ore deposits) and, using Nier's method, calculate the geologic ages of the rocks.

Apart from this technical achievement—which multiplied the usefulness of Nier's experimental technique a thousandfold—he determined the isotopic composition of the lead in present-day marine sediments, which were considered a natural average of the terrestrial crust. Assuming that all lead-206 and lead-207 were created by the decay of uranium—in other words, that neither was present when the Earth was created, which is certainly an approximation—he calculated a maximum age for the Earth: *5 billion years.* Since the oldest terrestrial rocks are 2.7 billion years old, he deduced that the Earth must have been formed between 2.7 and 5 billion years ago. Thus he "bracketed" the Earth's age. But that was only the first step.

On Harold Urey's suggestion, he then undertook an analysis of a series of very special rocks: meteorites. These are rocks that fall from the sky and whose extraterrestrial origin is proven. Patterson showed that the isotopic composition of lead in various meteorites followed a linear relationship (Figure 17). This relationship can be interpreted simply by assuming that the rocks were formed at the same time, with the same initial isotopic composition, and have evolved since then in different media with different uranium concentrations.

Patterson found an age of 4.55 billion years to be common to all the meteorites. Discovering a meteorite mineral (sulfide) that did not contain uranium, he determined the composition of "primordial" lead. Then he analyzed manganese nodules found on the ocean bottom and in marine sediments, which were assumed to represent the average of the terrestrial crust, and basalts from Hawaii, which were assumed to represent the Earth's mantle. He ascertained that their lead isotopic compositions placed them on the line he had deter-

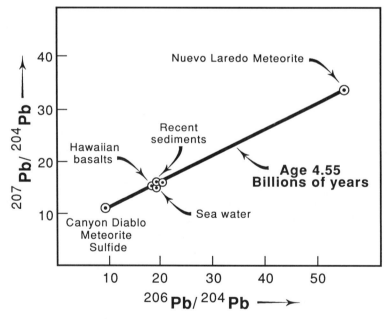

Figure 17 The Patterson lead isotope isochron showing that the Earth, iron meteorites (Canyon Diablo), and differentiated meteorites have the same age, 4.55 billion years.

mined for various meteorites and concluded from this that the meteorites and terrestrial reservoirs were formed at the same time, 4.55 billion years ago (Patterson, 1956). He was convinced that he had pinpointed the time of the formation of the planets in the solar system. The geologic calendar had finally found its January 1. Geologic time was not infinite.

The Geologic Calendar

Patterson's work is among the most important in all geology, but it should be placed in a more general scientific context: the development of geochronology based on long-lived radioactive isotopes. Just before the Second World War, the existence of other long-lived radioactive elements had been discovered: potassium-40, an isotope of potassium that decays to form argon-40, a well-known rare gas; rubidium-87, an isotope of a rare element in the Earth's crust that gives birth to strontium-87, another trace isotope; and carbon-14, which is produced high in the atmosphere and decays in a much shorter period of several thousand years (so that its use is restricted to archaeology).

In the postwar period all these methods—uranium-lead, potassium-argon, and rubidium-strontium—would be employed, thanks to the development in the United States of refined measuring techniques based on the use of the mass spectrometer and delicate microchemistry. Each method constituted an independent clock. Taken together, they could be used to verify each other, dating several minerals of the same rock by one clock or dating the same rocks by several clocks, to establish a true calendar of geologic time (see Figure 18). The various divisions made by stratigraphy were thus translated into absolute time by using a method that was both better and more reliable than Barrell's. In these efforts, laboratories in the United States, especially those at the Carnegie Institution in Washington, D.C., Caltech, MIT, and Columbia, led the way.

I will refer to this calendar later, but first I want to make two general comments. There seems to have been a considerable time interval—4 billion years—between the creation of the Earth and the appearance of the first abundant fossils during the Cambrian, and an interval of 550 million years between the first fossils and the appearance of Man. Yet 75 percent of all research in geology covers the last

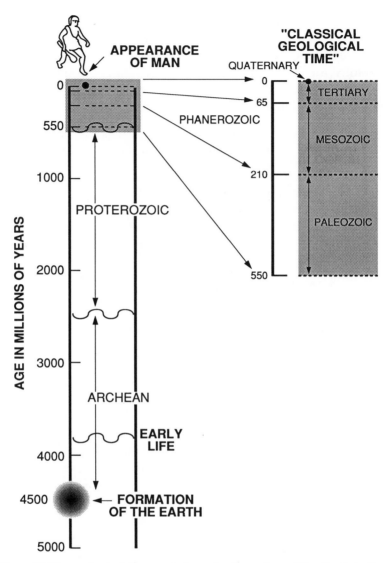

Figure 18 The geological time scale from the formation of the Earth to the present. The section (highlighted on the right) calibrated by stratigraphic periods represents only 550 million years. The appearance of Man is very recent.

200 million years and only 5 percent the 2.5 billion years of the Archaean and Proterozoic, an extraordinary example of the myopia of classical geology, whose choice of research topics has in no way corresponded to the relative abundance of the available rocky "documents."

This focus on more recent periods may also be a result of how the Earth works. Since Archaean terrains are rare at the surface of the Earth and young terrains are in the majority, the Earth must be a very active planet, forever destroying old rocks and creating new ones, eroding old mountains and building new ones. One fact has to be clearly understood. On the Earth there are no rocks that have an age of 4.5 billion years; the age of the Earth is determined by the "average values" found for red clay sediments and basalts. Meteorites, however, are real rocks with an age of 4.5 billion years. These differences simply reflect the fact that the rocks formed when the Earth was young have been destroyed and replaced by new ones, which is why the Earth is said to be a geologically active planet, while meteorites are pieces of planetary objects that are geologically dead. Such characteristics emphasize a fundamental aspect of the Earth, that it is a living planet, and explain Hutton's claim, "no vestige of a beginning." This was true on the scale of rocks or minerals, as Patterson showed, but it was not true on the atomic scale of isotopes.

4

Stones Falling from the Sky

According to Pliny the Elder, people have observed strange stones falling from the sky since antiquity. They cross the atmosphere, illuminating it with a vibrant glow, and smash into a thousand pieces upon impact with the Earth, leaving a record of that impact in the form of holes called craters. Some are very large and weigh several tons; others are smaller, and their weight is measured only in kilos. These stones that fall from the sky are called meteorites.

As one can well imagine, meteorites have intrigued people and attracted the curiosity of scientists ever since their discovery. At the end of the eighteenth century the question of their origin was the subject of impassioned and stormy debate. That it could be extraterrestrial was denied by most of the learned men of that time. Thus, Thomas Jefferson, a Southerner, could declare when a meteorite fall in New York had been described that it was easier for him to believe the Yankee scientists had lied than to admit that stones could fall from the sky! He thought the phenomenon absurd. The sky consisted of gas: how could stones, the very essence of solid objects, come from a gas? Only after a meticulous study of the Aigle meteorite by Jean Baptiste Biot of France and a careful report to the French Academy of Sciences was the extraterrestrial origin of meteorites accepted by the scientific community. It was not until the last thirty years, however, that the true nature of meteorites was well understood and their importance appreciated. It was necessary, in fact, to wait until progress in chemical analysis and the determination of their geologic age by radioactive dating made is possible to place them in a wider context. Today we know that they are messengers from the universe, or more exactly, witnesses to the primitive history of the solar system that, after a voyage through space and time, bring us decisive information about epoques whose traces have been almost totally erased from the Earth by the passage of time. Meteorites are in some sense witnesses to the origin of the solar system, to our origin.

The Age and Chemical Composition of Meteorites

As a result of Clair Patterson's work we know that meteorites are the oldest rocks that can be dated by radioactive methods and that they are as old as the Earth. But because Patterson's measurements were made using only the lead-lead method and on a restricted number of samples, the general applicability of his conclusions could be doubted. Since the Second World War and the technological advances ensuing from it, however, it has become possible to date a large variety of meteorites by the uranium-lead, rubidium-strontium, or potassium-argon methods. Patterson's initial conclusion that meteorites were formed 4.55 to 4.5 billion years ago has been confirmed completely (Minster et al., 1982; Tatsumoto et al., 1973; Turner, 1977).

The second observation that forcefully established the importance of meteorites is their chemical composition. Eighty percent of them have a chemical composition very similar to that of the solar corona (Wood, 1968). The luminous radiation emitted by the sun can be analyzed by decomposing the sun's white light into its principal components through a prism. Each ray of the optical spectrum can be attributed to a chemical element, which emits that characteristic radiation when excited in the hot atmosphere of the sun. The intensity of the radiation is a function of the abundance of the element. In this way, astronomers were able to determine the chemical composition of the sun; in the same way meteorites can be analyzed chemically in the laboratory.

Except for hydrogen and helium, which are the most abundant elements in the sun but whose gaseous and volatile character prevents them from being abundant in a piece of solid rock, the analyt-

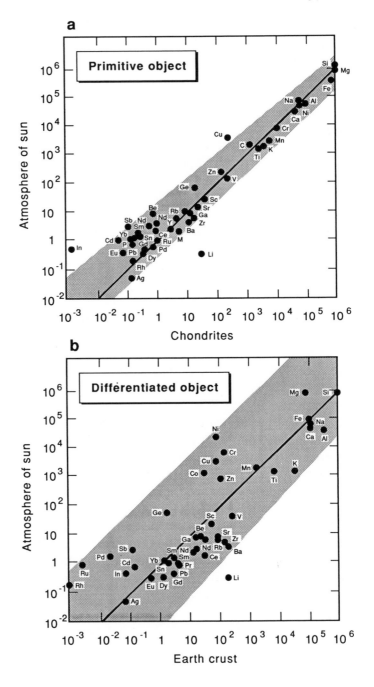

ical diagrams charting the chemical composition of the sun and of meteorites are remarkably similar (see Figure 19). Such similarities are very uncommon; if we were to compare the analysis of a terrestrial rock chosen at random—limestone, schist, granite, or basalt—with that of the sun, we would find little similarity between the two.

Meteorites thus appear to be unusual rocks. They are primeval, since they are as old as the Earth itself; they are primitive in chemical composition, since they resemble the sun, which contains 99 percent of the mass of the solar system and is no doubt very close to the "primitive cloud" in chemical composition. From this realization it is only a step to the conclusion that meteorites are samples of the primitive solid matter from which the Earth and the other planets were formed. For this reason, each meteorite that falls to Earth is treated with respect: it is given a name (that of the locality where it fell), it is recorded in catalogues like a precious art object, and it is preserved and shown in museums of natural history. At present several thousand samples have been indexed in this way.

Chondrites

Of each one hundred meteorites that fall to Earth, eighty are of the type called *chondrites,* which consist primarily of small silicated spheres called *chondrules.* Such structures are unknown among terrestrial rocks. If we disregard this particular characteristic and look at the mineralogy of chondrites, we see that they consist of silicates similar to those that form the terrestrial peridotites, to which has been added a very special mineral, pure iron. Particles of this pure iron are distributed among the silicated minerals and coexist with them.

Let's do a little experiment. We crush a chondrite and extract its iron particles with a magnet. Examined mineralogically and analyzed chemically, the residue bears an astonishing resemblance to a

Figure 19 Two diagrams showing a comparison of the chemical composition of the atmosphere of the sun with a) the chondritic meteorites; b) the Earth's crust. A perfect match would be represented by the straight line. It is clear that b shows a larger dispersion, while a fits the correlation line very well. This reflects the primitive character of a meteorite compared to the evolved character of terrestrial reservoirs.

crushed terrestrial peridotite. Potentially the agglomerated iron particles could constitute elements of the Earth's core. Carrying the analogy further, the proportion of the mass of free iron particles to silicates in chondrites is similar to that of the core and the mantle of the Earth. If we recall the two hypotheses for the Earth's formation, heterogeneous and homogeneous accretion, the idea that chondritic material represents the homogeneous solid from which the Earth was agglomerated and then differentiated seems to support the latter.

A simple scenario suggests itself: the terrestrial embryo accreted from chondritic material. Large-scale chemical differentiation took place, separating out iron and concentrating it in the core, and leaving the silicates to encircle it as a mantle. Here again we find the scenario of homogeneous accretion. The structure of chondrites, in which iron and silicates are intimately mixed, seems to offer a decisive argument in its favor. We understand why geologists have constant recourse to a chondritic model of the Earth and why the study of chondrites takes on additional interest. Chondrites constitute the primordial material of the solar system, transmitted intact through time. A detailed study of this material can teach us how it formed, how it agglomerated, and where it comes from.

Chondrites are clearly stones from Genesis sent by the heavens!

Differentiated Meteorites

Not all meteorites are chondrites. Some, although composed of silicates, contain no chondrules, and their chemical composition is closer to that of terrestrial rocks than to that of the sun. Since they are extraterrestrial in origin but contain no chondrules, they are called *achondrites*. The most common type of achondrite consists of pieces of basalt very similar to the basalts found on Earth as products of volcanic activity. But achondrites are not of terrestrial origin.

A determination of their age shows that basaltic achondrites solidified from molten lavas 4.55 billion years ago. Where? Somewhere in the universe on a planet that was undergoing intense volcanism at the dawn of geologic time. So volcanism is not a new or modern phenomenon, as the ancients thought, nor is it exclusively terrestrial.

Other meteorites whose composition is even more surprising, although familiar, also fall from the sky. These are *iron meteorites* or *siderites,* which consist of metallic iron (in fact, an iron-nickel alloy).

They are as hard as anything produced by modern metallurgy, so hard that ancient peoples used them to make weapons. Their similarity to terrestrial metallurgic products greatly facilitated their study, because all the experimental and theoretical information accumulated by metallurgists could be used. Microscopic examination of iron meteorites has shown that the vast majority are the product of the solidification of a metallic bath, that is, of molten iron.

The chemical composition of these meteorites of pure iron (and a little nickel) is quite different from that of the sun or of chondrites; it is just as different from that of basaltic achondrites. Yet their geologic age is 4.55 billion years, so they are also relics of the cosmos's archaic period. Achondrites and iron meteorites are very special meteorites: in some sense they are the "opposite" of chondrites. Whereas chondrites consist of a scattering of silicates and little pieces of iron, in the other meteorites pure iron and silicates are separated (little or no silicates in iron meteorites, no iron in achondrites). That is why we speak of *differentiated meteorites,* meteorites that have undergone an episode of chemical differentiation. They are less primitive and more evolved, in the planetary sense of the term, than chondrites. On this basis we can extend the scenario of chondritic homogeneous accretion (see Figure 20):

> *Stage 1:* Agglomeration and accretion of a well-mixed chondritic material that is locally heterogeneous but globally homogeneous.
> *Stage 2:* Melting phenomena in the interior of the planetary embryo, drawing the molten iron to the center to form the core while, near the surface, volcanic eruptions spurt out molten basaltic lavas.

The scenario increases in credibility when we remember that meteorites consisting of a block of metallic iron adhering to a block of stone rich in olivine (a complex silicate of magnesium and iron) also exist. In terms of this scenario these meteorites, called *siderolites,* could be samples of the core-mantle interface. The study of chondrites supports homogeneous accretion, as does that of differentiated meteorites, which also indicates that melting phenomena have played a large part in the process of differentiation of planetary bodies.

Meteorites thus seem to be irreplaceable witnesses to the birth of the solar system. When the solar system was being formed, planetary

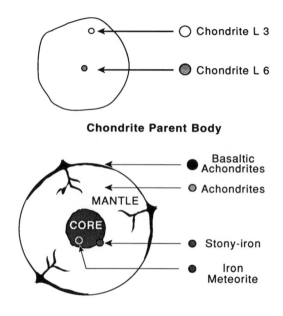

Chondrite Parent Body

Differentiated Parent Body

Figure 20 History of two types of meteorites, the chondrites and the differentiated meteorites.

bodies also formed. Some were primitive in composition, so that their iron particles and silicates remained closely mixed. Others differentiated themselves, forming a core of iron and displaying a volcanism at their surface suggestive of intense internal activity. Then these planetary bodies were fragmented and broken into a thousand pieces—pieces that were preserved in space from the ravages of time. Today, 4.5 billion years after the first episodes of their history, these rocks from Genesis are falling out of the sky. When they cross our atmosphere their exteriors melt and the most fragile of them break, without changing their interiors. When they hit the ground, the least hard fragment even more (see Figure 21).

Where do these stones come from? How has their celestial message been recorded and preserved? We will learn the answer in a little while, but for now let us continue to regard them as gifts from the sky.

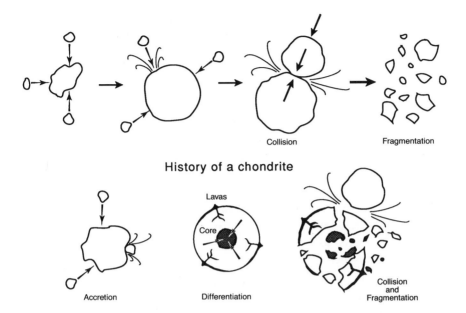

History of a chondrite

**History of a Basaltic Achondrite
or an Iron Meteorite**

Figure 21 Global histories of the meteorites showing accretion and fragmentation.

Iron and Silicates

Pure iron and silicates are the two essential components of the Earth, and they have been separated into two distinct and well-defined areas: the iron core on one hand, and the mostly silicate envelopes (mantle and crust) on the other. Differentiated meteorites also exhibit this iron-silicate separation, which thus appears to be one of the major processes—if not *the* major process—in the differentiation of planetary bodies.

In chondrites, on the contrary, we find pieces of iron and silicates closely mixed together in a continuous aggregation. But what is the exact relationship between the two components?

An essential property of iron is its ability to adapt to the chemical environment around it. Thus we say that it can assume several valence states. In an oxygen-rich (oxidizing) environment it is trivalent: it can chemically bond with three other elements. In an environment with a moderate free oxygen content, it is divalent. In an oxygen-poor (or hydrogen-rich) environment it cannot bond with other atoms but only with itself; it then forms metallic iron. Iron is the only abundant element that exhibits this strange, chameleonlike behavior.

Because meteorites contain metallic iron and terrestrial surface rocks do not, we know that the two environments were different when these two kinds of rock were formed. Thus, in nature iron is found in two forms: in the metallic state or bonded with oxygen (and, in the latter case, bound up with it in silicate compounds). Using metallurgical processes, it is possible to put oxidated iron in a reducing environment so that it breaks its bond with oxygen and becomes isolated as metallic iron.

In fact, however, in chemistry nothing is a hundred percent. Iron is never totally in the metallic state. Since there is always a little oxygen in the vicinity, there is always a little bit of oxidated iron. By measuring the ratio of reduced iron to oxidated iron, it is possible to determine the abundance of oxygen at the time of formation. This property of iron was used by Harold Urey and Harmon Craig in the 1950s, when they were at the University of Chicago, to classify chondrites. The proportion of metallic iron to iron bonded to silicates in each meteorite varies greatly. In certain chondrites all the iron is oxidated and there are no particles of free iron. These meteorites, which lack metallic iron, contain a lot of carbon and are thus called *carbonaceous* meteorites or C meteorites. Other meteorites that contain no iron in their silicates are called *enstatite* (enstatite is a mineral containing no iron) or E meteorites. Two groups between these two extremes show moderate degrees of oxidation. Urey and Craig established that the global iron content (the sum of the metallic iron and silicated iron) varies and distinguished a high iron trend (H) and a low iron trend (L).

Chondrites are thus grouped into four principal classes: E, H, L, and C, reflecting their formation in increasingly more oxygenated environments. To this grouping of differentiated meteorites and chondrites must be added a classification based on chondrite formation conditions.

Assuming that the Earth was formed from a homogeneous material of the chondrite type, we might ask where it would be found on a Urey-Craig diagram, and what the conditions during its formation were. Taking the core iron as reduced iron and the iron in mantle rocks as silicated iron, we can place the Earth between the H and E chondrites, that is, in an environment that is fairly poor in free oxygen but not totally deprived of it. Within our scheme of planetary differentiation from a chondritic material, we can imagine the two following extreme cases: a planetary body differentiated in very reducing conditions produces a large core of pure iron and a mantle totally lacking in iron; or an oxidated planetary body forms no core but is a homogeneous sphere very rich in iron. Thus we see how oxidation conditions can determine planetary structure. As a result of these observations it is not surprising to observe big differences in the chemical and physical properties of the different planets.

Gas and Dust

Any element and any chemical compound can exist in three states: the solid state, in which its atoms are rigidly connected; the liquid state, in which the intermolecular bonds are relatively loose; and finally the gaseous state, in which atoms and molecules are essentially free. At very low pressure only two states exist, solid and gaseous, and the passage from one to the other is sudden. Under these conditions ice *sublimates* into water vapor, and water vapor *condenses* into ice.

In the cosmos, as far as telescopes can reach into interstellar space, we know that matter exists in the two states of gas and dust. A gaseous atmosphere surrounds the solid Earth. In the solar system the relative proportion of gases and solids varies from planet to planet. The solid-gas problem is therefore at the heart of the chemical workings of planetary bodies. Naturally, the gaseous constituents and the solid ones are different elements. Iron and the silicates form the solids of the universe. Hydrogen, helium, nitrogen, oxygen, and like gases make up the gaseous portion. Some elements are gaseous; others are not. However, this distinction, which seems so clear and absolute, must be qualified. On a planet whose surface temperature is $-10°$ C water is present as ice, a solid. On a planet whose temperature is $500°$ C, water is found as vapor. What is true for water is true for all chemical compounds and elements. They are solid or gaseous according to the ambient temperature. For each compound and each element we can define a temperature at which it passes from the solid to the gaseous state: this is the temperature of *vaporization* or *volatilization.*

If we classify elements according to their vaporization temperature, we can derive a volatility scale. An element that vaporizes at a lower temperature than another is more volatile (nitrogen, for example, is thus more volatile than tungsten).

Very volatile elements: hydrogen, helium, argon, neon, xenon, nitrogen, carbon.
Volatile elements: indium, mercury, lead, sulfur.
Somewhat volatile elements: sodium, potassium, zinc.
Not very volatile elements: iron, magnesium, silicon.
Very refractory elements: aluminum, calcium, titanium, uranium, thorium.

Let us imagine a theoretical cosmos. At a given temperature, all chemical elements and compounds whose volatility is less than the ambient temperature are in the state of solid dust. All other elements are gaseous. The greater or lesser abundance of volatile elements in an agglomeration of cosmic dust depends on the formation and agglomeration temperature.

With this schema in mind, Ed Anders of the University of Chicago attempted to learn whether different chondrites had comparable concentrations of volatile elements. He discovered that the percentages of volatile elements in chondrites are extremely variable. Some, such as carbonaceous chondrites, are rich in volatile elements; others, like certain H, L, or E chondrites, are poor in volatiles.

In fact, when we compare the chemical composition of chondrites one by one with that of the sun, we realize that those that most closely resemble the sun are those that are richest in volatile elements. A deficiency of volatile elements (compared to the sun) is therefore an important characteristic in the evolution of cosmic material. Anders concluded that the thermal formation conditions of chondrites were very variable, from cold conditions for carbonaceous chondrites to warmer conditions for H, L, and E chondrites that are poor in volatiles. But to understand this phenomenon more completely we must look closely at the structure of chondrites.

Chondrites consist of two agglomerated parts assembled in a heterogeous way: chondrules, the spherical particles we have spoken of, and an intersticial cement that is called the matrix. The matrix consists of pieces of agglomerated minerals.

Edward Anders of the University of Chicago,
one of the masters of meteorite studies.

The relative proportion of chondrules to matrix varies. Some chondrites contain many almost perfect chondrules, whereas others contain few. Internal examination of the chondrules shows that they are the product of the cooling of a melted silicated liquid. The crystallization of minerals and free iron particles in the interior and the solidified crust that surrounds them suggest a huge number of independently solidified magma droplets. Chemical analysis of the chondrules shows, as one can imagine, their extreme lack of volatile elements.

The chondrules' matrices are much more variable in nature. Some were formed from minerals that crystallized at high temperatures, analogous in every way to those that form chondrules. Others are more heterogeneous and also contain compounds or minerals whose low-temperature origin is obvious. The matrices of carbonaceous chondrites thus contain complex carbonate molecules that a slight increase in heat inevitably destroys—clays, gypsum, carbonates, all minerals characteristic of the "cold" conditions found on Earth's surface. Analysis of chondrite matrices fully confirms the mineralogical observations and refines them by giving them a quantitative expression. The volatile element content varies according to the matrices: "cold" matrices are rich in volatiles; the others are not.

Anders's observations were beginning to be explained: the variation in volatile content depends not only on the proportion of chondrule to matrix but also on the volatile content of the matrices themselves. Some matrices are "cold," others are "hot."

At the beginning of time chondrules were created in this way during a hot phase; then, in a period of cooler temperatures, all the solids and dust in the region agglomerated. This scenario indicates that in the formation of the primitive solar system a first generation of dust formed into chondrules; then the temperature decreased and low-temperature dust formed in some places; later on, when the temperature was low enough, the agglomeration of dust and debris formed a meteoric body.

According to this description, carbonaceous chondrites appear to have been formed in peculiar conditions, since up to 5 percent of their weight is trapped water and rare gases that are found in much greater abundance than in ordinary chondrites. This led to the idea that carbonated meteorites are the hearts of defunct and burst comets, a hypothesis that has been substantiated by recent cometary missions, which have shown that comets are indeed cold objects whose cores are a mixture of silicates and ices. We must bring back a comet sample before we can say any more, but the recent flyby mission to Halley's comet has offered support for this idea.

The Metamorphism of Meteorites

Microscopic observation of the geometric relationship between chondrules and matrices complicates the problem even more. In some cases the contact observed under a microscope is sharp and clear, and the spherical chondrules seem to be embedded in a different substance. In other cases, on the contrary, the edges of the chondrules are "eaten up," and some minerals seem to be attached to both the chondrules and the matrices. In fact, it looks as if the chondrules and matrices, stuck together mechanically, have reacted chemically with each other in a secondary reaction. This is called chondritic *meta-*

morphism by analogy with the metamorphism of terrestrial rocks, which, through the action of internal heat, transforms a limestone into marble and a clay into schist.

The phenomenon of chondritic metamorphism shows clearly that after the agglomeration of the chondrite heating again took place, initiating mineralogical reactions. This phenomenon affects different types of E, H, L, and LL chondrites; some of these meteorites are strongly metamorphosed, others less so. Metamorphism is thus a phenomenon unrelated to the Urey-Craig classification. This metamorphism, which is a secondary reheating, is capable of expelling very volatile chemical elements into space. John Wasson of the University of California at Los Angeles used this possibility as an argument to rebut Anders's theory. He advanced the idea that the variable distribution of volatile elements simply reflected the secondary phenomenon of reheating, not a primary phenomenon that had taken place during the accretion of the meteorite. Wasson's interpretation (1974) implies that a secondary reheating took place after the cold agglomeration of meteorites. What was the energy source for this reheating? Did it take place in the center of the parent bodies? Or, on the contrary, did it take place at the periphery, following an irradiation due to the intense activity of the very hot, primitive sun? At present there are no satisfactory answers to these questions which, we can sense, harbor fundamental problems for our understanding of the formation of planets and for the reconstruction of the first moments of the solar system. As for the distribution of the volatile elements, what determines it? The initial cooling off? The heliocentric position of the meteorite in the primitive nebula and its distance from the sun in particular? Or should we see only evidence of subsequent metamorphism in this distribution?

Let us try to enlarge our understanding of these difficult problems by attempting to place all these events in a precise chronological sequence.

The Detailed Chronology of the History of Meteorites

Meteorites were formed 4.55 billion years ago. This information is critical, since it chronologically locates not only the history of the Earth, but also the entire development of the solar system. It is insufficient, however, because we need to know how long it took to form the planetary bodies and in particular the meteorites or parent bodies. One hundred million years? One million years? A thousand years? Scenarios for the formation of the planetary objects would be radically different, depending on the response to this question. The alternation of cold and hot episodes would be perceived differently. How can we obtain such precision from our chronological methods for events that took place more than 4 billion years ago?

The answer to this difficult question first came from the University of California at Berkeley, and more particularly from John Reynolds. In 1960 Reynolds discovered that the isotopic composition of a rare gas, xenon, taken from the Richardton meteorite, was distinctly abnormal. The isotope of mass 129 was abnormally abundant. To Reynolds this was not just any old isotope; it was potentially the disintegration product of iodine-129, since lead-206 is the ultimate disintegration product of the uranium-238 chain—with one small difference, however: iodine-129 does not exist in nature today!

Various astrophysical theories claim that this isotope existed at the beginning of the solar system but that because of its very short disintegration period (17 million years) it has been inexorably destroyed. How, then, to prove its past existence and its disintegration into xenon-129? .

Reynolds and Peter Jeffery (Jeffery and Reynolds, 1961) had recourse to an elegant method of degassing in steps. Xenon is buried in minerals, but when they are heated it breaks loose according to a law that reflects the manner in which it is bonded to them. When a meteorite is heated, xenon does not degas below a certain temperature, but once the critical temperature is reached, it degasses suddenly.

Then Reynolds and Jeffery had the idea of submitting their meteorite to the flux of a reactor to provoke an artificial nuclear reaction and transform a portion of the iodine-127, a stable isotope present today, into xenon-128. Degassing their irradiated meteorite in steps, they ascertained that all the isotopes of xenon degassed at a temperature of 1,200°C. At a slightly higher temperature, however, they were pleased to see that the meteorite was still giving off xenon, but this time in a very unusual isotope. Instead of the nine usual isotopes, this high-temperature xenon consisted of only two; xenon-129 and xenon-128, created artificially, which certainly proved that the excess xenon-129 came from the same mineral "site" as the iodine.

His hypothesis corroborated, Reynolds looked for anomalies of xenon-129 in other meteorites and found them, in greater or lesser quantities. Measuring the amount of iodine-127 in all of them and assuming that the iodine-129/iodine-127 ratio was the same at the beginning of time, he was able to establish a relative chronology among the various meteorites.

The astonishing result, which he announced with his students Chuck Hohenberg and Frank Podosek, was that all the meteorites were formed in the extremely "short" time interval—short for cosmochemistry, that is—of 20 million years (Hohenberg et al., 1967). Thus, 4.55 billion years ago, the solid bodies of the solar system formed over a period of only 20 million years.

Reynolds had to wait almost twenty years to see his results confirmed by more traditional methods, such as the rubidium-strontium. Early on, results of the rubidium-strontium method seemed to invalidate the iodine-xenon interpretation. They suggested that the formation time for all the meteorites seemed to be closer to 150 million years than to 15 million. A "long" chronology apparently replaced the "short" chronology. Since the rubidium-strontium method seemed more reliable than the slightly exotic iodine-xenon method, those conclusions appeared to be valid. But a few years ago work done in our laboratory at the University of Paris explained the contradiction and reestablished agreement between the rubidium-strontium and iodine-xenon methods (Minster et al., 1982). Indeed the short chronology is right!

The initial formation of *all* meteorites seems to have taken place in 10 or even 5 million years, exactly 4.552 billion years ago. This formation led to the fixing of the rubidium-strontium ratio of the various meteorites. Since rubidium is a volatile element there were different thermal scenarios during these primitive phenomena, like those suggested by Anders. But 50 to 70 million years after the initial formation, reheating caused metamorphism in the chondrites. When we determine the age of the minerals in a metamorphosed meteorite, we are determining the age of the metamorphism.

Within this scenario, the precise chronology of differentiated meteorites allows us to say that the extraterrestrial volcanism of meteoritic basalts took place 20 to 30 million years after the formation of the chondrites, which shows that the differentiated meteorites were heated faster and hotter than the chondritic bodies.

If we compare the metamorphism of the chondrites and the formation of the basaltic achondrites (see Figure 22), we must conclude
that the cause of the reheating is to be sought in the interior of these
pseudo-planets. But what was the heat source?

We will have to return to this enigma later.

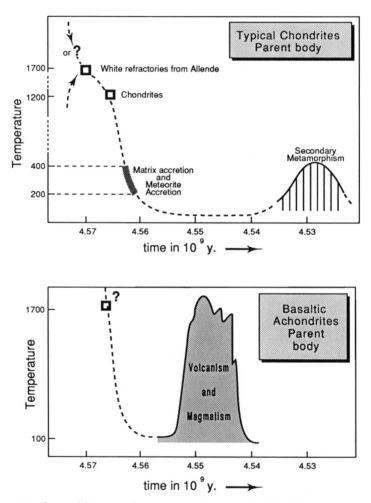

Figure 22 Thermal history of two meteorite parent bodies. Top: a chondrite
parent body that has undergone metamorphism. Bottom: a basaltic
achondrite parent body with a reheating episode creating early volcanism.

The Condensation Model

All the observations supported by the characteristics of element volatility gave birth to a theoretical model for the formation of meteorites and, more generally, of planetary objects. This is the *condensation model,* which must be attributed to Harold Urey (1952), but which was systematically developed by H. C. Lord in 1965 and then "rediscovered" and popularized by John Larimer and Ed Anders (1967) at the University of Chicago on the one hand, and Larry Grossman (1972), who was then a student at Yale, on the other.

As a model for the future solar system they proposed a *hot* gaseous nebula whose chemical composition is identical to that of the sun today (remember that the sun contains 99 percent of the mass of the entire solar system). Being hot, the nebula emitted rays—light—into space, thus losing some of its heat. It cooled off. Eventually the temperature at which certain compounds are no longer stable in the gaseous state was reached, and those compounds condensed not into liquids but into solids, for the pressure in space was very low. The nebula became filled with solid dust grains. These grains formed larger and larger solid objects by accumulating at first into meteorites and then later, if there were enough of them, into planets.

The question that Lord, Larimer and Anders, and Grossman posed was that of the precise chemical and mineralogical composition of these solid condensation grains, which are therefore called condensates. To arrive at an answer they had recourse to the calculation of *chemical equilibria.* For each temperature they considered all possible chemical equilibria among all gaseous and solid materials capable of forming in a mixture of solar composition and determined which compounds are in the solid state in these conditions. They thus obtained a series of chemical compounds that are deposited successively when the temperature of a "solar gas" falls. This is called the *condensation sequence* (see Figure 23).

The first compounds that condense at 1,300°C are oxides rich in titanium, aluminum and calcium. These kinds of compounds are used today as bricks in high temperature industrial furnaces. At about 1,050°C metallic iron condenses massively (this is an atmosphere rich in hydrogen); at 950°C the first silicate, in this case the silicate of magnesium called olivine, then other silicates of magnesium and iron that are called pyroxenes; at about 800°C silicates with weak struc-

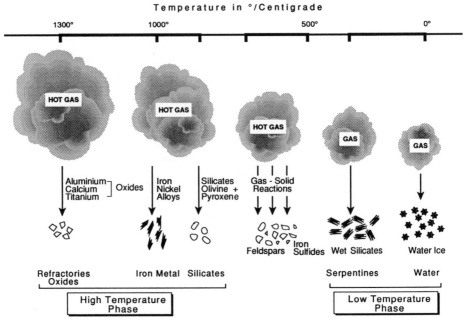

Figure 23 A schematic diagram illustrating the condensation sequence of solids from a gradually cooling hot nebula (the sequence has been greatly simplified).

tures, plagioclase feldspars and iron sulfide; at even lower temperatures a silicate containing water, serpentine, a kind of "olivine-clay." Finally, at 0°C, water condenses to ice.

One cannot help but be struck by the nature of the minerals that appear in this condensation sequence. They are not random compounds, occurring by chance among the several thousand naturally existing compounds. Pure iron (alloyed with a little nickel), the first abundant condensate, is the component of the terrestrial core and of iron meteorites. It is the only abundant metal that condenses to the metallic state, and it is also the only metal that is found in abundance in the cosmos.

As for silicon, it doesn't condense to a metal but forms compounds. The silicates of magnesium and iron that are called olivine and pyroxene, which form the first silicate condensates, are essential ingredients of the terrestrial mantle, of ultrabasic rocks, and also of chondrites.

The next condensate, feldspar, joins with the pyroxenes to produce basalt, the ubiquitous rock that is found on the ocean floor as well as in meteorites. Thus, with the four principal solid condensates it is possible to "fabricate" chondrites, iron meteorites, achondrites, and the components of the Earth's interior.

We can also explain the formation of carbonaceous chondrites by assuming that the low-temperature compounds with clays and water were mixed with the high-temperature compounds. Condensation calculations allow us to quantify and correlate the chemical composition, and especially the volatile content, mineralogical composition, and formation temperature, in a more complete way.

We can easily understand why this theory was an immediate success with the cosmochemists. Far from elucidating the question of homogeneous or heterogeneous accreation, it simply renders both possible a priori. If solid grains accrete and join together as they condense, the first to form is an iron core that is surrounded by a silicate mantle and then by compounds rich in water. This is the heterogeneous accretion model. If, on the other hand, accretion and agglomeration of the grains take place only after condensation has ended, solid bodies with a homogeneous chemical and mineralogical composition are formed. Both scenarios seem to be intact.

Not completely, however. If the correct scenario is that of homogeneous accretion, once the grains are condensed and are not agglomerating they can react with gas and produce new minerals. Thus metallic iron can react with gaseous hydrogen sulfide to produce iron pyrite, a mineral that is well known in meteorites. If, on the contrary, the iron is locked into a solid agglomeration, iron pyrite can form only by a secondary reaction farther from the center of the solid body. It is the same for the other compounds. Close examination of meteorites seems to indicate that all these "reactives" are original and not secondary, which seems to support the homogeneous accretion theory. The very existence of meteorites such as chondrites with a globally homogeneous structure seems to point in the same direction and strengthen the homogeneous accretion scenario.

The condensation sequence is therefore an important step in our understanding of the formation of the solar system. If we look at it more closely, however, we can see that we have skipped over the first condensation products, the oxides rich in titanium, aluminum, and calcium, whose importance we have never established at any time, either in meteorites or on Earth. As the first condensates in the calculation, they seem to play no role. Do they exist?

Of course, it is always possible to assume that these relatively low-abundance compounds were later destroyed during secondary processes, either in the interiors of planetary bodies or by reacting with the gas in the nebula. But such a hypothesis seems a little ad hoc, invented for the needs of the case, to explain a defect or weakness of the model.

Is Allende Planetology's Rosetta Stone?

Jean-François Champollion was able to decipher Egyptian hieroglyphics only as a result of an exhaustive study of the Rosetta Stone, on which a message was written in hieroglyphics, demotic, and Greek. This stone has become the emblem of those who are trying to reconstruct the past by reading stones or who hope to find the one that will allow them to reconstruct a puzzle for which they have only scattered pieces. In 1969 Mireille Christophe Michel-Lévy, a mineralogist at the University of Paris who was using microscopic examination to study carbonaceous meteorites, discovered in one of them, Virgano, some white minerals consisting of oxides of titanium, aluminum, and calcium (Michel-Lévy, 1969). Several months later, Ursula Marvin, at the Smithsonian Astrophysical Observatory in Cambridge, Massachusetts, confirmed this observation on another carbonaceous meteorite (Marvin et al., 1970). These observations increased the credibility of the condensation model, since it was now predictive as well as explanatory and synthetic. Condensates rich in titanium, aluminum, and calcium had been predicted by the calculations; now they had been found in nature! This discovery, however, provoked only a modest interest. It was not until two years later that it received the kind of attention it deserved.

On February 8, 1969, a two-ton meteorite fell near the Mexican village of Pueblito de Allende. It was an amazing stroke of luck. This meteorite was of the carbonaceous type, an exceptionally interesting and extremely rare one—before this we had a total of only a few dozen kilos of it. Studied immediately by Ursula Marvin and John Wood, it proved to be extremely rich in "reactive" inclusions of oxides of titanium, aluminum, and calcium that would come to be called *Allende white inclusions.*

Pursuing these investigations, John Wood and Larry Grossman thought they could assert that since the white inclusions were surrounded by free iron particles and then by olivine and pyroxene, they had in effect observed the condensation sequence, as the calculations had predicted. Grossman analyzed the trace elements contained in the various phases of Allende and showed that the white inclusions are the poorest in volatile elements ever observed. The iodine-xenon chronology made by Frank Podosek of Washington University in Saint Louis set the age of these white inclusions as the oldest ever found for any rocky object. Jerry Wasserburg's group at Caltech also determined the primitive character of Allende by the rubidium-strontium method. This was confirmed recently by several groups, including ours, using the uranium-lead and lead isotope techniques.

Everything fitted together perfectly. Allende held the mysteries of Creation within it. The white inclusions were the first solid grains to be formed in the solar system. The condensation sequence is the key to explaining the formation of planets and therefore of our Earth.

5

The Planetological Adventure

When Neil Armstrong stepped onto lunar soil on July 20, 1969, he did not yet know that he was opening a decade of planetary exploration whose scientific harvest would totally revolutionize our knowledge of the solar system.

Previously, we knew the planets only from snapshots taken through terrestrial telescopes whose essential characteristic was not clarity. Today we possess a complete set of detailed photographs of the surfaces of Mercury; Venus; Mars and its two satellites; the moon, of course; Jupiter and its four large satellites, Io, Europa, Ganymede, and Callisto; Saturn and its largest satellites, Titan, Rhea, and Dione; Uranus and its satellites; and now Neptune and its satellites. For many of these planets we have precise geophysical measurements (of the magnetic field or the field of gravity), and for three of them–the moon, Mars, and Venus—we have chemical analyses of the materials that constitute their surfaces.

Today our knowledge of the Earth is no longer isolated. We can situate our planet among a collection of similar objects, homologous to it but, as we shall see, different from it. How could we speak of the Earth's formation or earliest evolution without referring to the planetological context the space missions have made it possible to reconstruct? Before attempting to draw the conclusions that will illuminate our subject, I will retrace this systematic exploration, because each step of the planetological adventure had its own discoveries, and we have lived this adventure day by day, mission after mission, marveling at each one of them.

First Step: Lunar Exploration

When President John F. Kennedy, responding to the Soviet Sputnik challenge, made a moon landing a major NASA objective, Harold Urey had already fixed this objective for the young planetological community. Some scientists are visionaries.

The moon has always been a source of fascination for human beings, but for a long time it remained only the same pale disk, since, as a result of the astronomical phenomenon of resonance, it always presented the same face to terrestrial observers. Since the Apollo missions, which were followed by the Soviet Luna missions, the moon has become for us a little "planet" whose topography, internal structure, surface rocks, and geologic history are known to us. The accumulation of all this information became a ten-year scientific adventure for researchers of very different origins, none of whom had any particular a priori knowledge of the moon, but who quickly congregated around the Apollo project and formed the embryo of today's large planetological community (Taylor, 1982).

Harold Urey, an ardent advocate of planetary exploration and Nobel laureate.

The moon's surface consists of two distinct elements: the waterless seas or *maria* (singular *mare*, Latin for "sea"), the dark, flat, low-lying plains, and the highlands, the light, rough, and deeply undulating regions that surround them. The maria have very distinct shapes and, in fact, together resemble a collection of different-sized circles superimposed upon each other. Only the visible face of the moon possesses these maria; its hidden face contains only mountains (an observation that has still not been explained). The seas, like the mountains, are riddled with craters. We have had all this information since the time of Apollo 9 and Apollo 10, which placed a satellite in lunar orbit. We also knew that there was an accumulation of dense material under the maria that showed itself in positive anomalies of the field of gravity called *mascons* (mass concentrations).

It was only with Apollo 11, however, which landed in the center of the Sea of Tranquillity, and the return of the first rock samples that the lunar adventure really began. The first samples picked up by Neil Armstrong in the immediate vicinity of the lunar module arrived in Houston in August 1969. The possibility of microbial lunar life meant that the samples had to be kept isolated in totally aseptic conditions in an impervious place quarantined from all exterior contact, the Lunar Receiving Laboratory. In a fever of excitement that we can well imagine, a few chosen scientists undertook the first analyses. They soon realized that the rocks taken from the soil around the lunar module were pieces of basaltic volcanic lava, very similar to the basalts we find on Earth that make up the majority of volcanic lavas and the ocean floor. The lunar soil itself, a gray powder that uniformly covered the surface, consisted of debris crushed into a million pieces. The first chemical analyses did not indicate anything spectacular, with the possible exception of the low iron content of the rocks compared to terrestrial basalts. When the quarantine was over and careful biological examination had revealed no lunar microbes, the samples were distributed parsimoniously to the best rock analysis laboratories in the world. Several months later NASA organized the first "lunar geology" conference in Houston, which caused the sacred term *selenology* quickly to fall into disuse. This semantic shift reflected a real transition: a qualitative change had taken place in the method of studying the moon, a clean break with the methods of astronomers in favor of the methods of terrestrial geologists.

Back in Houston, the first results were about to be presented in a highly charged atmosphere. The most eagerly awaited results were those concerning the age of the lunar samples. Were they as old as meteorites, as many imagined? Were they much more recent, as others hardly dared hope?

The ages furnished by Jerry Wasserburg's group at Caltech using the rubidium-strontium method, by Mitsunobu Tatsumoto of the U.S. Geological Survey in Denver using the uranium-lead method, and by the young Grenville Turner of Sheffield, England (whose proposal had been rejected by the English selection committee but was rescued by NASA) using the potassium-argon method all coincided. The rocks from the Sea of Tranquillity were 3.8 billion years old. But strangely enough, the soil or lunar dust showed an age of 4.55 billion years, the same as that of the Earth and of meteorites.

The principle of stratigraphy—an upper, younger layer covering an older layer—thus appeared to be violated. Tom Gold of Cornell University hastily concluded that the soil must therefore be "extralunar" and consist of debris of meteoric origin. Unfortunately, chemical and isotopic analyses weakened this theory and led to the conclusion that the soil and the adjacent basalt were identical in composition. If the soil is 4.55 billion years old, it is because it is a mixture, an average of all the surface rocks mixed together. Further analysis of the rocks confirmed the preliminary results. The most spectacular of the chemical results concerned the elements of the rare earth family: unlike terrestrial basalts, those from the moon were marked by an extraordinary deficit in one of them, europium. Europium is preferentially associated with plagioclase feldspar. On this basis John Wood of the Smithsonian Astrophysical Observatory concluded that the light-colored highlands must consist of a rock called anorthosite, which is composed almost entirely of plagioclase. More detailed analysis of the various types of rock showed the existence of a special type, breccia, which consists of agglomerated fragments held together by a mineral cement. The multiple origins of these fragments suggested the phenomena of secondary impact and agglomeration, features already known to exist among meteorites. Their discovery on the moon made it possible to understand the origin of meteoritic breccia, which had probably been formed by impact with a parent body. This idea was reinforced by the discovery that some of the breccias had been formed by the agglutination of some of the lunar soil itself.

All these observations, combined with an examination of the close-up photographs taken on the lunar surface, made it possible to put together a very coherent scenario for the principal elements of lunar geology at this early stage. The craters are not volcanic craters, as some had thought, but impact craters created by an intense meteoritic bombardment. This incessant bombardment has broken, chopped up, and pulverized into dust the basalts that form the subbasement of the maria. The basalts seemed to have poured out in the form of unusually fluid lava flows (Mutch, 1970). But what are the sources of these lavas? We see no volcanoes or calderas or fissures. Perhaps this very fluid lava had also covered up the openings through which it came to the surface?

We have not yet mentioned the organic products in this "hit parade" of new discoveries. On July 30 this startling declaration was sent out from Houston: "There is organic matter on the moon!" Mass spectrometers had detected it. Speculation immediately ran wild. Some newspapers headlined: "Life on the Moon!" But it soon became necessary to come down to earth, so to speak. Analysis showed that the organic products were only fuel emitted by the reactors of the lunar module which, when lighted, contaminated the surrounding rocks . . . ephemeral glory for the promoters of exobiology!

A seismograph left on the moon registered a new phenomenon, "moonquakes." The shape of their signals is very different from those of earthquakes, and seismologists announced that the moon's interior must certainly be less rigid than the Earth's. But they had only one registering station at their disposal and therefore could neither locate the moonquakes nor study the trajectories of the vibrations and their perturbations as we do on Earth. We would have to wait for subsequent missions to learn any more.

Negative observations are also extremely important in planetary exploration missions. The first Houston conference made it possible to gather together what had been learned in this direction. As we have already noted, there is no life on the moon, no great volcanic cones, no water. Although a variety of detection methods was used to search for it, water is definitely absent. Geologically speaking, there are no great faults as on Earth, either, nor are there elongated mountain ranges. The magnetic field is nonexistent. The planet's geology appears to be dominated by only two phenomena: volcanism and the impact of projectiles of various sizes.

Apollo 12, following soon after Apollo 11, did not change our knowledge fundamentally. The landing took place in the Ocean of Storms. The astronauts, liberated from some of the apprehension of the first voyage, received permission to move away from the lunar module, and as a result their collection of rocks was larger. Rocks from this site are also basaltic. Their texture is sometimes that of lava, sometimes that of breccia. They are 3.25 billion years old. A light soil, as thick as that at the Apollo 11 site, covers the ground. Continuous observations made on the ground as well as in orbit confirmed those made by the first mission.

After Apollo 12 we began to experience the lunar program in a different way. From enthralled spectators we became diligent but impatient participants whose commitment sublimated our enthusiasm. The American lunar samples were not the first to arrive in Paris. Apollo 13 never reached its destination, and before Apollo 14 could bring us the long-awaited rocks, the Soviets had sent an unmanned vehicle called Luna 16 to the Sea of Fertility and brought back rock samples. Several grams had been offered to France, and our laboratory at the University of Paris received some of them. They were swiftly followed by samples from Apollo 14, then by those from Apollo 15, Luna 20, and Apollo 17, and finally, much later by those from Luna 24. NASA's bet on us was fruitful, and we remained "on board" through Apollo 16 and Apollo 17 and were included in the Preliminary Mission Team that made the first analyses immediately after the mission and before the "general distribution."

This is a digression, but I mention these rock samples because my interest in cosmic problems probably began with these analyses. Luna 16 had allowed us to increase our knowledge of the lunar maria. Apollo 14 and the landing on the Fra Mauro crater, and then Apollo 16, Apollo 17, and Luna 20 provided us with more information on the lunar mountains.

As we have already noted, the lunar mountains contain many more craters than the maria, so their surface has a very chaotic and tortured appearance. This caused difficulties in landing and meant that the seas were given priority for exploration. We quickly realized that the mountain rocks are very different from those of the seas. As John Wood had predicted, the light-colored rock of the mountains consisted of plagioclase: it was anorthosite.

Long experience in studying terrestrial rocks had taught us that this rock is not simply a solidification of melted magma, as are basalts. It implies a mechanical separation from the magmatic bath that produced it. This separation can take place because the plagioclase mineral is less dense than the silicated bath and therefore floats and accumulates on its surface. We thus hypothesized that at a certain time the lunar surface melted and the plagioclase crystals floated up to form a superficial crust.

Systematic study of the samples revealed the existence of certain rocks that were less rich in plagioclase but had a very special chemical composition characterized by their richness in potassium, phosphorus, and the rare earths. They are called the KREEP (K for potassium, REE for rare earths, P for phosphorus) rocks. But the common characteristic of all these rocks is that they are breccias. None of them is intact; they have all been subjected to meteoric bombardment. They appear to be very old—4.4 billion years—much older than the volcanic rocks from the seas. Now we understand why the mountains are more cratered than the seas: since they are older, they have been bombarded much longer.

Starting with Apollo 16 a small lunar vehicle was used to investigate the mountains, which allowed the astronauts to really explore the region, and the essential role of meteorite impacts in planetary geology was confirmed. Impacts create craters. When these craters are deep, they allow access to the moon's interior and reveal the KREEP rocks under the anorthosites. When they are shallow, they break up the surface rocks, creating the dusty soil strewn with debris and pulverized rocks. But the effect of the impacts is not only destructive. Impacts agglomerate the rocks and "glue" them together, creating breccias consisting of pieces of rock of different origins. Large impacts can reheat adjacent terrains and produce melting in the interior of the planet, which sets off magmatic phenomena at the same time.

As we can see, lunar geology opened a whole new chapter for the geologic sciences: the study of impacts and their effects. The first trails in this direction had been blazed by such daring pioneers as Edward Chao and Eugene Shoemaker of the U.S. Geological Survey and Robert Dietz of Arizona State University ten years before the lunar exploration, but it must be said that no one paid much attention to their work. Lunar exploration put their studies back into the mainstream and, as we shall see, that was only the beginning (Mutch, 1973).

More than five years of study of the various lunar rock formations, their chemical analysis, and their dating made it possible to develop a scenario for the moon's history that is accepted by the majority of scientists today. Let me summarize it.

About 4.4 to 4.5 billion years ago, part of the moon melted. Except for a very thin surface crust that had cooled, most of the interior was liquid and formed a gigantic magmatic chamber like those found under volcanoes. This state is often thought of as a vast *ocean of magma.* Since the enormous reservoir of lava was losing heat through its roof, its temperature was decreasing. Eventually it reached the temperature at which some crystals take solid form. The homogeneous bath was then transformed into a liquid-solid mixture, but the mixture was not stable and it began to settle out. If the crystals were lighter than the bath, they floated on the surface, forming a crystalline scum. If they were heavier, they fell to the floor of the magmatic chamber. In the moon's case the plagioclase crystals began to float, separating out an anorthosite crust, while the olivine crystals fell to the bottom. In this way the plagioclase mountains formed the primitive crust of the moon. As this double differentiation process continued, the chemical elements that took no part in the formation of plagioclase or olivine remained in a bath whose volume became smaller and smaller. Their concentration in the liquid increased, as the salinity of a body of water increases when the water is cooled and ice forms on it. Among the chemical elements that "salted" the residual lunar bath were potassium, phosphorus, and a whole series of minor elements, among which were the rare earths. The famous KREEP rocks of which we have spoken were formed in this way under the anorthosite layer. This process ended 4.3 billion years ago, when the moon was almost totally solidified. Its surface, however, continued to be bombarded by projectiles, some of which were of considerable size (several dozen kilometers in diameter) and these pocked its surface with craters of various sizes. When a projectile hits a planet it digs a crater; at the same time, the shock of the impact releases a certain amount of heat to the planet, as a slap heats the cheek of the person who has been slapped. We can understand how this simple mechanism caused the moon's interior to maintain a certain temperature. Another influence—created by the radioactivity of the lunar rocks, or rather, by the radioactivity of the uranium and thorium in the lunar rocks—was superimposed on this one. It would take nearly a billion years for these cumulative effects to make themselves felt. Then 3.8 billion years ago a new phenomenon developed: volcanism began to appear in the vast circular basins created by the giant meteorites. The moon's interior, heated in situ by radioactivity and ex situ by impacts, began to melt. The magma thus formed easily found its way to the surface, for the roof rocks had been cracked by the impacts. Gigantic volcanic flows began to fill the depressions and create lunar seas. Then the phenomenon ceased, and the density and intensity of the impacts decreased. Since the accumulated heat had been evacuated by the volcanism, the interior cooled off. For eternity.

Today there is only a small central zone 200 kilometers in diameter to which seismologists attribute a plastic quality, the last trace of the hot magmatic interior. For the last 3.2 billion years the moon has been a dead "planet." Only meteorite impacts still disturb the stillness of its surface.

Seismographs placed in several spots—unfortunately all on the same side—have made it possible to locate the sources of the moonquakes and their origin. They take place at a depth of about 700 kilometers, and their cause is simply the attraction that the Earth exercises on the moon. They are therefore related to the lunar tides. The study of moonquakes has made it possible to determine the internal structure of the moon using the same techniques developed and tested for the Earth. Unlike the Earth, the moon has no dense core, which is not surprising when we remember that its average density, 3.5 grams per cubic centimeter, is almost the same as that of its surface rocks. Its internal structure consists of a mantle surrounded by a crust.

This internal structure is more surprising when examined from a chemical point of view. The lack of a core would normally mean that iron, instead of segregating itself in the center, as on Earth, remained dispersed among the rocks, just as it is found in chondrites. The iron concentration of the lunar mantle should be analogous to that of the chondrites and greater than that of terrestrial rocks. But, surprise: the iron content of the lunar basalts, products of mantle melting, is lower than that of terrestrial basalts. We are forced to conclude that the moon is poorer in iron than the Earth. What became of its iron?

The moon has neither atmosphere nor ocean. One could therefore suppose that the gases that on Earth form these envelopes have remained in the moon's interior and have not been expelled or pulled toward the surface. Analysis of lunar rocks shows, however, that the moon's interior is poor in nitrogen, carbon dioxide, and water. Systematic analysis of all the chemical elements shows that the volatile elements such as lead, zinc, and mercury are much less abundant than they are in terrestrial rocks. Poor in iron, the moon is also poor in volatile compounds. This dual lack seems strange if we remember that iron is among the first elements to condense out, while the volatile compounds are among the last.

A chemical comparison of the moon and the Earth quite naturally leads to questions about the relationship between the two. Why does the Earth have the moon as a satellite?

The first hypothesis was the capture hypothesis. As the moon was traveling through the universe, it was attracted by the gravitational pull of the Earth and finally captured by it. In this case the moon would be a "foreign body," a sort of prisoner or victim of the attractive power of the Earth. The study of this hypothesis by terrestrial astrodynamicists has shown it to present serious difficulties. Without going into the details, let me simply say that calculations show that the moon would have been so perturbed by its capture that it would have exploded.

Then it was proposed that the moon was derived from the Earth through fission. Since the density of the moon is similar to that of the terrestrial mantle, it was thought that the moon could have been separated from the Earth *after* the differentiation of the core. The moon would thus be a piece of the terrestrial mantle. Our planet's scar from this surgery would be the gigantic Pacific Ocean. This hypothesis also ran into serious difficulties. What force could have torn off a piece of the Earth? Centrifugal force, of course: the force that allows us to throw a rock a long distance by using a sling. Calculations show, however, that for such a phenomenon to take place the Earth-moon system would have to be turning a lot faster than it is in fact turning. Moreover, a detailed study of their respective chemical compositions has shown that the moon and the terrestrial mantle have very different concentrations of iron and the volatile elements.

A third hypothesis is that the moon was formed like the Earth, through the accretion of dust in the terrestrial environment. Neither captive nor daughter of the Earth, the moon would then be its little sister. Even then, however, it is hard to conceive how the moon could in that case be less rich in iron, a very nonvolatile element if ever there was one, and poorer in volatiles than the Earth.

Mariner 9 and the Discovery of Mars

Lunar exploration was not yet completed when NASA launched a new probe, this time toward Mars. There was no question of landing there, even less of bringing back rock samples, but simply of launching into an orbit around Mars a satellite whose task was to photograph the surface of the red planet with a resolution of 200 meters. Compared to the Apollo program, this mission was relatively modest. It furnished a series of data, however, that marked a turning point in the exploration of the solar system. What some expected to be only a photographic mission turned out to have a crucial scientific importance.

The photographs transmitted by Mariner 9 immediately showed us that the soil of Mars, like that of the moon, is pocked with impact craters. Some regions are more cratered than others and therefore older than others, but the whole planet is affected. The analogy with the moon stops there (Mutch et al., 1976).

The surface relief on Mars is much more varied and contorted than that on the moon. It includes magnificent shield volcanoes like those of Hawaii but much more majestic. Mars boasts the largest volcano in the solar system, Olympus Mons, which is 25 kilometers high, 600 kilometers wide, and has a crater 20 kilometers in diameter. Martian volcanoes are found either singly or in groups. Some in the Tharsis region are laid out like terrestrial volcanic ranges. There are also real stream canyons with tributaries branching out into veins like those on Earth. The largest of these canyons, Coprates, is 3,000 kilometers long. But its valleys are dry: no trace of running water, no trace of rain, no trace of transport—they resemble gigantic desert wadis. The analogy does not end there. Mars, like the Earth, has polar caps. Its poles are covered with white ice. But what kind of ice? Since the atmosphere of Mars consists of carbon dioxide, is it carbon dioxide snow? Or, more simply, is it water ice? At this point the first mission had not yet answered this question.

But the mission continued and the data—the photographs—came flooding back. A formidable dust storm darkened the planet's atmosphere and made it impossible to take photos for several days at a time. The dust storms' complement was quickly found: near the poles lay sand dunes similar to those of the Sahara or Gobi deserts.

Canyons, volcanoes, glaciers, deserts: the Martian landscape looked much more familiar to us than the lunar landscape. The satellite continued inexorably sending its photographs back to Earth. It eventually became possible to draw a "geologic" map of the planet, which showed it to be divided into two provinces separated by an equatorial circle.

The southern hemisphere, riddled with about as many craters as the moon, contains the river canyons; the northern hemisphere contains many fewer craters and in some places is covered by great plains formed by lava flows that are completely analogous to the lunar seas. But the most intriguing observation was the progressive melting of the polar caps in "summer," leaving remnant caps and an extremely spectacular series of stratified deposits. Spectroscopic analyses made in orbit identified the nature of the melting polar caps: frozen carbon dioxide (dry ice).

As the orbital examination continued, we realized that the on-board magnetometers indicated the absence of a magnetic field. Spectrometers analyzed the atmosphere of Mars precisely: although capable of giving rise to those obstructive sand storms, it is in no way comparable to that of the Earth, either in its pressure or in its composition. Pressure at soil level is one fifth of terrestrial atmospheric pressure, and the composition of the Martian atmosphere is dominated by carbon dioxide, followed by nitrogen. There is no oxygen.

At the end of this astonishing mission, the great unanswered question was about water. The study of the temperature and pressure conditions on the Martian surface showed that water could not exist there in the liquid state. Ice and vapor were the only two stable states possible. How, then, had the fluvial canyons been formed? Did the remnant polar caps consist of ice? These questions sparked debate among specialists without yielding any clear answers. In addition, no "little green men" had been found, and no direct or indirect evidence of life. It is true that with a resolution of 50 meters, even elephants would have looked rather fuzzy.

NASA decided to send another mission to learn more. The object of this mission, called Viking, was to land two automatic vehicles on Mars—a task at which the Soviets had failed several times when their machines crashed into the planet. Almost the entire mission was to be devoted to the search for life on Mars, so the vast majority of the experiments scheduled were in the field that is still called exobiology. Unfortunately, since the experiments were badly conceived and ill-prepared, it took a long time to decipher the results, and they were ambiguous although mixed with many new and, as one might expect, exciting developments.

As far as we know, there is no life on Mars. The automatic analysis of the powdery Martian soil, strewn with debris that looks like volcanic rock, did not provide any clearer answers than an analysis of a little Saharan sand would provide extraterrestrials interested in the history of the Earth. The most important results were obtained by the mass spectrometer that analyzed the composition of the atmosphere. This experiment by Alfred Nier—the same man who analyzed the isotopic composition of uranium ores in the late 1930s—confirmed the abundances of the principal components but showed in particular an excess of xenon-129 (the isotope produced by the extinct radioactivity of iodine-129, as discovered by Reynolds in a meteorite) larger than that in the terrestrial atmosphere. Finally, although no really original new measurements were made, the Viking mission allowed specialists to agree that Mars had once had a meteorology with rainfall and flowing water, which explained the presence of fluvial channels. The atmosphere, which was dense enough to have allowed such a meteorology, is now stored in the soil and in the glacial caps. Mars's atmosphere had a history, and Mars had been subjected to climatic variations.

We can imagine the speculation on the origin of these variations: variations in the inclination of the axis of rotation? variations in the ellipticity of the orbit due to the attraction of Jupiter? Since we have not yet been able to agree on the causes of terrestrial climatic variations, it is easy to understand the degree of uncertainty that exists for Mars! This mission was only a beginning. Mars is an exciting planet about which we still have much to learn.

Mariner 10 and Mercury

Since Mercury is so near the sun, it is difficult for astronomers to photograph. The pictures sent back by the Mariner 10 probe were really our first concrete images of this planet. The general impression is simple. Only an expert can distinguish a photograph of a region of Mercury from one of a lunar sea at first glance. Lava flows and craters of various dimensions are combined as they are on the moon. In addition there are fractures that could have been caused by the cooling of the planet. The high density of the craters indicates that, like the moon, Mercury has been dead for several billion years (that is, it has no geologic activity) but was formerly the site of abundant volcanic activity. In addition, it has no atmosphere, which eliminates erosion by water or wind. The only surprise in the exploration of this planet was in finding a magnetic field, dipolar like ours although more than a thousand times weaker. Thus, Mercury, whose density is 5.5 (which is similar to the Earth's 5.3) but whose size is almost three times smaller does not have its materials compressed. It consists of a large, dense core, no doubt made of iron, surrounded by a rather thin silicate mantle. The origin of its magnetic field should be sought in the iron core, as it is on Earth. Is the field weaker because Mercury's rotation is less rapid or because, since it is smaller, Mercury has less energy?

Venus, or the Triumph of Radar

Venus's size and density are similar to those of the Earth, but there are several differences. First of all, Venus spins very slowly in the direction opposite to its orbital rotation, one orbit lasting 240 terrestrial days. Most important, it is covered by a very thick atmosphere whose mass is ninety times that of the Earth's. As a result, the surface pressure on Venus is ninety times atmospheric pressure, which is the equivalent of the pressure that obtains at a depth of 1,000 meters in our terrestrial oceans. The atmosphere is so dense that it traps the sun's rays and creates a greenhouse effect on the surface, raising the surface temperature to 470°C, making any life impossible and creating extremely severe "climatic" conditions. Temperature and pressure conditions on the Venusian surface are similar to those that cause metamorphism in solid rocks in the depths of the Earth. The most conspicuous feature of Venus is its atmosphere, and Mariner 10 gave us the first precise picture of it. As the French amateur astronomer Boyer observed, this atmosphere turns much faster than the planet itself. Its rotation takes four days, which creates a real spinning-top motion around the planet. This behavior is strange and certainly very different from the terrestrial case, in which the atmosphere is pulled along by the planet. Although various models have been proposed, the explanation for this atmospheric rotation remains to be found. The atmosphere consists of carbon dioxide and a little nitrogen and closely resembles that of Mars in composition, although it is far more voluminous. Does it contain water or not? Mariner 10 photographed abundant clouds and an active atmospheric circulation. What is this meteorology? What are the clouds made of?

Since the thick atmosphere prevents any view of the surface, even after the Mariner 10 mission we still did not know what it might look like. Several Soviet and one American mission have been sent to Venus since then. The Soviets succeeded in penetrating the atmosphere and landing on the surface. "Landing" on Venus is easier than on Mars, because the high density of the atmosphere makes it possible to use a parachute. While crossing through the atmosphere, the landing modules were able to measure its composition and, upon landing, to give indications of the nature of the surface materials. The atmosphere contains water, but in small quantity, and this water remains at high altitude. It takes part in reactions that make hydrochloric acid and, especially, sulfuric acid. These acids form heavy clouds. Venus's meteorology is therefore highly corrosive—not a water meteorology but an acid one!—yet the Soviet and American landers crossed it without being dissolved. Once on the ground and subjected to extremely harsh conditions—470°C and a pressure of 100 atmospheres—the teleguided landers and their electronics resisted these conditions for several hours. This Soviet technical achievement made it possible to do a chemical analysis of the surface rocks, which revealed that some of them resemble terrestrial granites while others are rather basaltic. Photographs taken on the ground show dense rocks sitting on what appears to be soil or a carpet of dust. The American Pioneer Venus mission used radar to pierce the atmosphere and make a topographic map of Venus's surface. This map was completed by using measurements made by the giant radar net of Arecibo observatory in Puerto Rico.

The surface of Venus is rather flat. More than 60 percent of the planet has reliefs whose heights are less than 1,000 meters. Venus has several "mountains" or continents whose altitudes are more than 2,000 meters above the plains, but these mountains represent only 5 percent of the surface. Two regions among them have been studied in detail: Maxwell Montes, which rises to a height of 11,000 meters, and the Lakshmi Plateau, which is 3,000 meters high but completely flat on top. The plateau is bordered on both the north and the south by a range of higher mountains. This complex is called Ishtar after the Assyrian and Babylonian goddess of love and fertility. A much lower plateau in the southern hemisphere is called Aphrodite after the Greek goddess of love.

The existence of real continents inspired the participants in the Pioneer Venus mission to look closely at Venus's gravitational field by measuring the altitude variations of the satellite. As on Earth, the reliefs are compensated, that is, they obey Archimedes's principle. The reliefs consist of materials whose densities differ from those of the plains, just as the density of the terrestrial continents differs from that of the ocean floor. This observation reinforces the results of the rock analyses done in situ by the Soviets. Venus strangely resembles Earth, but it has no magnetic field. Closer observation of the radar-photos has led certain astrogeologists to assert that the valleys and escarpments could be ridges or faults. The Arecibo radar even seems to have detected a volcanic eruption. These observations await confirmation, but they are completely plausible.

On the other hand, numerous ringlike structures, interpreted as impact craters, have been detected, and a preliminary statistical study has established that the surface of Venus is slightly more cratered than the terrestrial surface but much less so than the moon, Mercury, or Mars. Thus the "evening star," of which we knew almost nothing, is revealed as our true sister planet.

The Giant Planets and the Voyager Missions

There is no way to make a prize list of the planetary missions, for each one wrote a new chapter in the development of our understanding of the universe. If emotion were allowed to count, however, I think the Voyager mission would have to be placed on the same level as the moon landing. Not only were planets located at distances on the order of a billion kilometers flown past by spacecraft launched by human beings, but the pale little points on the telescope photos, which were the only view we then had of their satellites, were transformed into a series of clean, clear images, thanks to which we can not only recognize them but also study and describe their "geology."

Before Voyager, we were acquainted with five planetoids; today we can recognize almost thirty. Now it is possible to speak of comparative planetology and to locate our Earth among a numerous group of objects.

Voyager brought us three principal sources of information: a better knowledge of the giant planets, a complete description of the structure of the solid objects (the rings) that gravitate around them, and finally, a set of exceptional data on their satellites.

Jupiter and Saturn

Jupiter is a giant planet, the largest in the solar system. It is three hundred times larger than the Earth, but its mass is only 3.18 times that of Earth, which gives it a density of 1.33 (as we have seen, that of the Earth is 5.3). This remarkable lightness is a reflection of its composition. The Earth and, more generally, the telluric (inner) planets are solid objects consisting of a mixture of iron and silicates surrounded by a little gaseous atmosphere. Jupiter consists essentially of hydrogen and a little helium. Its chemical composition is very similar to that of the sun. When we realize that, we must turn the question around: why then is its density as high as 1.3, greater than that of water, when we know that hydrogen and helium are extremely light gases? These gases, which the little planets were not able to retain, although they very probably formed the essential part of the protosolar cloud, *were* retained by Jupiter and Saturn because of their large masses. At the same time, this mass strongly compressed the materials situated toward the center of these planets, increasing their density. Thus hydrogen and helium, which are in the gaseous state on the surface of Jupiter are liquid toward its center.

The Voyager mission made it possible to determine the internal structure of Jupiter by using simple methods whose efficacity had been verified in the terrestrial case—Newton's laws of gravitation. When a body is rotating, each of its elements is subject to two opposing forces: the force of gravitational attraction, which tends to pull it toward the center of the body, and centrifugal force, which tends to expel it toward the exterior. The shape of the body and the distribution of masses in its interior reflect the equilibrium existing between these two forces. Because of this, all spinning spherical bodies tend to be flattened at the poles and to bulge at the equator. The bulge is larger if the mass is distributed uniformly in the planet. If it is concentrated toward the center, forming a core, the bulge is small. The equatorial bulge for Jupiter is only 6 percent. Knowing the amount of this bulge and the density of the planet, we were able to calculate that Jupiter should have a dense core probably consisting of ice and rocky materials; in brief, it is a kind of Earth.

The second fundamental observation made by Voyager is that Jupiter emits twice as much energy as it receives from the sun. Therefore Jupiter must have an internal source of energy. Is this source internal nuclear reactions, as in the stars? The mass of Jupiter is too small to reach or even approach the necessary temperatures. Without doubt the heat source results from the gravitational attraction that allowed the planet to agglomerate. This process is a good illustration of the physical principle of the conversion of different forms of energy. Potential energy is transformed into heat energy, the attraction that exists among particles causes them to collide, and, as a result of the collision, to be heated, as the skin is heated when it is rubbed vigorously.

This discovery had various consequences. First, it meant that we could calculate a thermal profile for Jupiter. This led to the conclusion that the temperatures in the center are 20,000°C to 30,000°C, which is ten times higher than those in the center of the Earth but one hundred times lower than those in the center of a small star. What we can deduce from this is that conditions are such that neither helium nor hydrogen can be found there in the solid state. The central core consisting of rocks and ice is probably surrounded by a liquid "mantle." Only the twenty kilometers nearest the surface are gaseous. The second consequence, whose meaning will soon become clear, is that since the phenomenon of accretion was much more important in the past, the heat emitted must also have been greater; at its birth Jupiter must have been in this respect a "little (nonnuclear!) sun," shining and shooting its rays into the cosmic space of its neighborhood.

Observations of Saturn lead to very similar conclusions, as much for its internal structure as for its thermal regime, even if the latter is complicated by the curious phenomenon of helium rain. Saturn actually has a helium meteorology, which transfers mass and energy to its interior from its surface. Helium rains there!

If the internal structure and the chemical composition are basic observations, the dynamics of the Jovian atmosphere are incontestably its most important phenomenon. There is a banded, zonal circulation, with whorls, twists, and whirlwinds within each band, reminding us that the circulation is extremely violent. Hot plumes arising from the interior of the planet create spots on the surface, of which the Great Red Spot is the best known and the most intriguing. Friedrich Busse of the University of California attributes this feature to the existence of convection currents in enclosed cylinders, upon which are superimposed transverse jets that pierce the calm zonal circulation. In any case, the study of Jovian meteorology will permit us to build planetary circulation models, one of whose objects will be to explain the terrestrial or Venusian circulation as well as that of Jupiter or Saturn.

Many mysteries still remain. The most obvious is the color of the atmosphere: what chemical compound gives Jupiter's atmosphere its russet color and makes its large spot red?

The Rings of Saturn, Jupiter, and Uranus

We were already well acquainted with Saturn's rings. Since Jean Dominique Cassini's time we have known that Saturn's aureola of small solid objects or dust had a ring structure separated by voids. But we knew no such thing for Jupiter. What a surprise for the Voyager observers when they realized that, like Saturn, Jupiter also possessed a ring system! At the same time, they discovered that Uranus was also surrounded by rings. The rings that a few years ago had seemed to be specific to Saturn are today considered a completely normal feature of a giant planet.

The rings consist of a myriad of small rocky fragments that are assembled into a disk whose width is a million times larger than its thickness (imagine a razor blade a thousand times thinner that those we use). This disk rotates in the equatorial plane of the planet. How can such a disk be stable? First of all, it is clear that the disk is rotating; otherwise the formidable gravitational attraction the mother planet exerts on it would cause it to fall into the planet. Each rock, each particle that is part of the disk, is turning around the planet. It is therefore subject to two forces: the attractive force of the planet and centrifugal force, which tends to thrust it away. The trajectory that it follows is defined by the equilibrium between the two. Thus the rings consist of an infinity of objects turning at great speed, each on a defined orbit. The mechanics of this are not perfect, however, and collisions occur. These collisions can break two projectiles into pieces or simply displace them in space. In both cases the collisions tend to separate the two projectiles laterally and thus to spread out the ring. As we can see, through this simple collision process a system of a few objects located in the same orbit can spread out into a thinner and thinner ring. But how did these rings form? Why don't the telluric planets have any?

The answer is not yet totally clear, but it is possible that in the nebula that surrounded Jupiter and Saturn before their condensation, the exterior gases, which were still abundant and dense, allowed rocky satellites to condense. In the near vicinity these satellites were numerous and began to hit each other and be fragmented, thus creating a little belt of rocks. The temperature elevation of the central planet after its contraction drove away the gas, letting the little projectiles continue their fragmentation and evolve into rings. This is one possible scenario.

Jupiter's Satellites

Around Jupiter, the giant of the giant planets, a real miniature solar system has formed. More than fifteen satellites, all situated in the equatorial plane, gravitate around it. Four of these satellites are of particular interest. They are the four that Galileo was able to see through his telescope: Io, Europa, Ganymede, and Callisto. Their dimensions are similar to those of our moon, and their densities diminish with their distance from Jupiter. The first two are primarily rocky; the latter two consist of ice.

Io's surface is rocky. Voyager photographs show that it is covered by a series of calderas of volcanic origin but that it has very few meteor craters. The volcanic calderas unleash lava flows whose morphology resembles those of Hawaii or the Martian volcanoes. By a happy chance, the Voyager spacecraft witnessed a volcanic eruption on Io. The erupting volcano emitted projection plumes that formed an umbrella-shaped sheaf. Everything indicates that intense volcanic activity has taken place there and is still occurring. Its cause seems to be a tidal effect exerted by Jupiter. Internal friction could initiate the melting of rocks and set off their expulsion as lava. But what is the nature of these rocks, which appear red or yellow in the photographs? Spectroscopy during the Voyager mission showed the presence of sulfur. Is this volcanism sulfuric or is it silicated volcanism with a high sulfur content? The question remains open; we will have to wait for other missions to learn the answer.

The surfaces of the other satellites are covered with ice. For Europa the ice is only a layer, since its density of 3 indicates that its interior is certainly rocky. Ganymede and Callisto, on the other hand, are made almost completely of ice and their surfaces are pocked with craters. On top of these numerous features great interrelated bands whose nature is unknown are superimposed. They are immense lava trails, but lavas of ice. On these satellites there must be volcanoes spitting out not melted lava but liquid water, which flows away before freezing into ice. Should we call them volcanoes or fountains? Europa, unlike its two icy sisters, has only a few craters but many trails. Does this indicate that its rocky interior caused such intense internal activity that the flows of icy water destroyed the traces of craterization? This is another question for the future.

These are, however, very curious planets: the craters are as numerous as those elsewhere, the surface features look like bunches of tangled strings, and the volcanoes are fountains!

The Satellites of Saturn

Saturn's satellites are quite similar to Jupiter's Ganymede but on a much larger scale. Tethys, Enceladus, and Dione, whose surfaces we have been able to photograph with good resolution, show ice structures with impact craters, great fractures, and real ice flows, attesting to the existence of water volcanoes—in brief, all that we have already seen, in a multiplicity that confirms its generality. But one satellite, Titan, is very unusual.

Titan is the only satellite in the solar system that has an atmosphere. This quality in itself justified deviating Voyager 1's course toward this interesting satellite with a radius of 2,300 kilometers in order to study it more closely. Unfortunately the photos were not as clear as had been hoped. The atmosphere, a uniform cloud, offered the cameras no window, and the surface remained invisible. We had to be content with indirect measurements made by radio and infrared spectrometry, which nevertheless produced substantial results.

Titan's atmosphere consists of nitrogen and methane, argon and hydrogen. Numerous hydrocarbons such as ethylene and acetylene are added to this mixture. On the surface, where the temperature is $-175°C$, methane is liquid. Titan may be covered by a methane ocean that produces a meterology with evaporation, wind, and rain, but the clouds and rain are made of methane. Carbon, which on Venus or Mars exists as carbon dioxide gas, here exists as hydrocarbon because of the lack of oxygen on Titan's surface. The interior of the planet may consist of a mixture of rock and ice that totally traps the oxygen, or the oxygen could be kept prisoner by the extremely low temperature of Titan, creating a veritable sea of "oil." If we landed on Titan to explore it, our vehicles would not lack fuel, but they would lack an agent of combustion because there is no oxygen. We would be reduced to navigating by sail on a sea of petroleum!

After Saturn, Voyager passed near Uranus, where it again found rings and a group of satellites. The photographs are quite faint but much better than anything we had before, and they reveal more than ten new satellites. Heading into the corner of the solar system Voyager passed near Neptune, and everything it found there was new—new sets of satellites, new observations (we knew of two satellites, Triton and Nereide, from astronomical observation, but now we know of six more)—a quite complete set of new planetary systems. It is still too early to comment extensively.

Certainly the exploration of the universe presents us with many variations on the common planetary theme: the Earth, our Earth, is an object like all the others, and like all the others, unique among its kind.

What planetary exploration has clearly demonstrated is the great difference between the amount of information we possess about the moon and what we know about the other planets, a difference due to the fact that we have actual samples only from the moon. Our laboratory techniques are so sophisticated, so superior to any remote analysis, that the amount of information extracted from sample analysis has no remote equivalent. The lesson is obvious. If we want to progress further in our understanding of the planets, we will have to build and send round-trip sampling missions to Mars and Venus, and eventually to the other planets. In this way we will learn a great deal about the origin of the solar system. The past twenty years have been wonderful, and the direction of future progress is clear.

6

From Newton to Mendeleyev

After our ride across the solar system to the rapid rhythm of Soviet-American competition, we need to take a breath, to survey and organize all the information we have gathered.

Planetary exploration costs a great deal of money and, as a result, the presentation of its discoveries has often favored its spectacular or "photographic" aspects, to the detriment of the fundamental information this unfinished quest has already brought us. This situation, inspired partly by the media style of the present day but also by the need to impress the public as quickly as possible—in order to make the best possible case for further funding—has greatly irritated many scientists, who have questioned the profitability of the space program. The Apollo program cost more than thirty billion dollars over five years. Landing an unmanned spacecraft on Mars (or Venus) to take samples and return to Earth would cost about four billion dollars.

I have no desire to determine the informational cost of space exploration. Rather, I want, more modestly, to put the planetological adventure back into the context of the studious and patient accumulation of data, which remains when the hoopla of the daily news has been forgotten. Like most scientific progress, planetary exploration has not increased our knowledge of the planets from zero to infinity. It has, however, radically changed our approach to acquiring this information. Let us retrace the path together.

Kepler's Solar System

We will begin our inquiry with a review of the classical astronomical model of the solar system. By summarizing the facts that form the traditional body of knowledge, built on centuries of astronomical observations and calculations, we can begin our journey from a solid, well-established base and measure the road covered, at least to the extent that it is measurable, in tangible, well-identified, and synthesizable discoveries:

- The planets revolve around the sun in *elliptical* orbits that are almost *circular.*
- Far from being oriented in all directions in space, the orbits are all located in a *plane* and therefore define a *disk,* which is perpendicular to the sun's axis of rotation.
- The planets' motion in their orbits follows unchanging rhythms. As they approach the sun, they accelerate. As they move away from it, they decelerate.
- The planets' periods of rotation depend on their distance from the sun: the farther away they are, the more slowly they revolve around it.
- The planets all revolve in their orbits in the same direction, that of the sun's rotation. It is also the direction in which the planets spin on their axes (except for Venus and Uranus), and their axes of rotation are almost perpendicular to the plane of the ecliptic (the plane on which the planets move).

These are the laws of celestial mechanics discovered by Kepler and explained by Newton.

But that is not all. In their distance from the sun the planets obey Bode's law. Broadly speaking, it says that each planet is twice as far from the sun as its nearest interior neighbor. Expressed in "astronomical units" equivalent to the distance from the sun to the Earth, this law is extremely precise and has only one "exception": between Mars and Jupiter, contrary to Bode's observation, instead of a single planet we find a myriad of small solid objects, the asteroids. Moreover, around the planets—especially the giant planets—revolve systems of satellites whose motion seems to mimic that of the planets around the sun, following a local version of Bode's law.

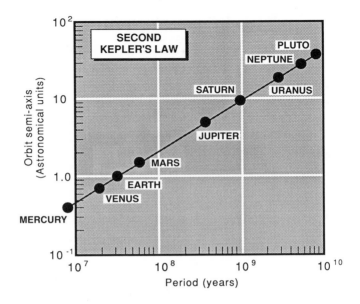

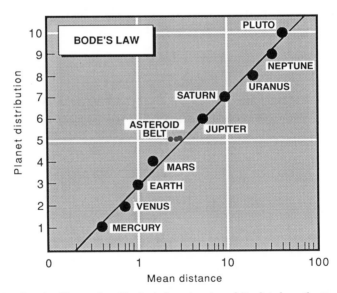

Figure 24 Graphs illustrating Kepler's law (top) and Bode's law (bottom), which show that planets are not randomly distributed but obey precise laws.

So the solar system appears to work like a gigantic well-regulated, well-oiled clock whose unchanging movement takes place according to strict rules (see Figure 24). How did such organization come into being? How did such a gigantic system organize itself so perfectly and so meticulously? If we can speak of Nature's laws anywhere, it is here. The mechanics of the solar system have been used for nearly two hundred years as an example of the perfection of these laws, which make it possible to predict the future from a knowledge of the past.

In Newton's time the prime mover of celestial mechanics was thought to be God. That explanation did not satisfy Pierre Simon de Laplace, who, when questioned by Napoleon on the existence of God remarked, "I have no need for that hypothesis." Yet Laplace also noted that such a "perfect" system could only have been created as a whole, a single entity. Simple calculations would show, he said, that a system consisting of objects with a variety of origins thrown together by chance would have none of the observed regularity of the solar system. That is true: the present solar system is the end product of a history that is common to all the planets. Looking for the origin of the various planets is a global, unitary exercise, and speaking about theories on the origin of the solar system thus has a universal significance. From this point of view, the Earth is simply one planet among others. Its formation can be understood only within the general framework of the origin of the solar system.

The Two Families of Theories on the Origin of the Solar System

For a long time the only object of theories about the solar system was to explain the formation of the planets and the dynamic regularities that we have just briefly reviewed. The exercise suggested to theoreticians was clear: Imagine an initial system that would evolve spontaneously into a planetary system respecting the laws of Newtonian mechanics. Without going back as far as the ancient Greeks or Egyptians, or even to Descartes's *vortex,* we can say that the two types of theories on the origin of the solar system can be traced to Buffon (1749) on the one hand and to Kant (1755) and Laplace (1796) on the other.

According to Buffon, at a certain point a *catastrophe* took place in the universe. A comet (which in those days was thought to have the characteristics of a true star) crashed into the sun. This collision caused the sun to eject a plume of matter that, in cooling and condensing, gave birth to the planets, which were distributed like a string of beads along a thread.

For Kant and Laplace, who independently proposed similar theories before 1800, the formation of the solar system—the sun and the planets—took place without the intervention of exterior forces. The sun formed through the contraction of a gaseous, rotating nebula. The nebula rapidly took the shape of a disk with a bulge or ball at its center. The ball began to grow, and its speed of rotation increased with its size. The centrifugal force acting on the ball also increased. From time to time rings of matter were left behind by this proto-sun, and the compaction of these residual rings produced the planets.

Today most scientists agree that the nebula theory is the most probable—with one difficulty, however, which has been the subject of passionate debate: the sun contains 99 percent of the mass in the

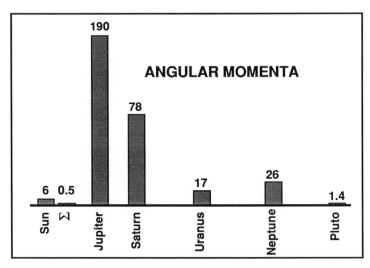

Figure 25 The distribution of angular momentum among different components of the solar system. Σ represents the total of the inner planets. This graph clearly illustrates the curious distribution of such parameters, which are still unexplained. Where is the angular momentum of the sun?

solar system but only 2 percent of the total angular momentum (moment of inertia times rotation speed), the majority of which is in the planets (see Figure 25). In other words, the sun turns too slowly for its mass. What happened to its angular momentum? This is still a great mystery.

The Accretion of the Planets

Until recently astronomers and physicists agreed that planetary accretion was caused by gravitational collapse. The cloud of nebular gas, changed into dust by cooling, would have been concentrated in some places. In those places the force of attraction among all the particles would have been greater than the thermal agitation (whose tendency would have been to disperse the particles), and the cloud would have contracted suddenly in a sort of implosion. The formation of a planet was thus a kind of catastrophe, a brutal avalanche occurring in a very brief period of time. In short, a planet formed much like a star.

After 1940, however, isolated from all scientific contact with those on the other side of the Iron Curtain, the Soviet school led by O. Y. Schmidt started to develop a completely different scenario. According to Schmidt, one of the fundamental points in the dynamics of the solar system is the *quasi-circular orbits* of the various planets. If the planets had been formed by gravitational contraction, he said, they would move in random elliptical orbits. From this observation, Schmidt and his students Levin and U. S. Safronov developed the *mathematical theory of progressive accretion* (Schmidt, 1944; Safronov, 1969) illustrated in Figure 26.

According to this theory the solid particles accrete first into marbles, the marbles into balls, the balls into larger balls, and so on. The small balls made of rocks, which are the embryos of planetary objects, are called *planetesimals.* As the process continues, the number of objects decreases and the proportion of large objects increases. How does the accretion process come about? All the objects in the solar system are in motion and revolve around the sun in elliptical orbits. Each solid body or planetesimal has its own trajectory and its own velocity. When two planetesimals meet, a variety of things can occur: if they are the same size, they collide with each other and either bounce off in different directions, hit each other so hard that

The Russian theorist O. Y. Schmidt.

they break up into a myriad of smaller pieces that also travel around in space, or (and this is the most attractive hypothesis to those who are trying to understand the formation of planets) adhere together to produce a larger ball. If they are of different sizes, in general the smaller is attracted to and adheres to the larger, whose size increases as a result. This is what happens when a meteorite falls to Earth. In other words, the formation of planets is like a gigantic snowball fight. The balls bounce off, break apart, or stick together, but in the end they are rolled up into one enormous ball, a planet-ball that has gathered up all the snowflakes in the surrounding area.

The ball has to attain a rather large size, more than 2,000 kilometers in diameter, in order to become plastic and rounded under the combined effects of rotation and gravitation. Phobos, a small satellite of Mars, has a diameter of 30 kilometers and is shaped like a big potato. The moon, which has a diameter of 3,400 kilometers, is spherical, as are the other satellites and planets. Mimas and Enceladus, satellites of Saturn, have diameters of only 400 kilometers yet are spherical. But they are made of ice, which is much more malleable and plastic than rocks.

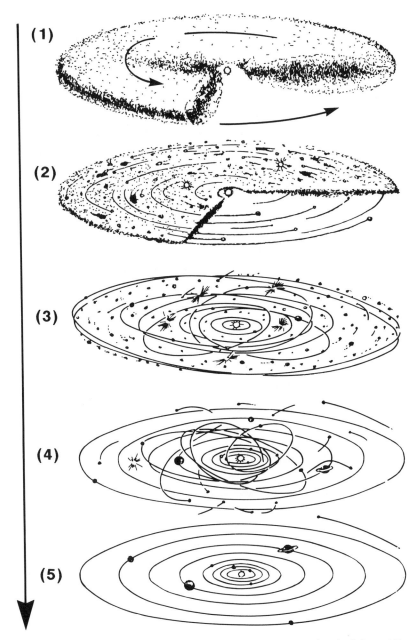

Figure 26 Schmidt's progressive accretion scenario: (1) a cloud of dust; (2) dust mixture with small embryos; (3) planetesimals; (4) protoplanets undergoing shock; (5) planet formation.

The Lunar Craters

Western scientists did not pay much attention to the Soviet theory, at least before the exploration of the moon.

When Neil Armstrong and Edwin ("Buzz") Aldrin set foot on the moon on July 20, 1969, they were confronted with a desolate landscape; their feet sank into a loose soil of gray dust filled with rocky fragments. In the module Michael Collins was frenetically photographing the moon's surface from all angles. These views—firsthand and photographic—revealed the same reality: the moon's surface is riddled with craters whose dimensions vary from a few meters to 60 kilometers in diameter.

Impacts have extracted, pulverized, and torn off rock fragments and volcanic dust. A detailed study of the resulting craters revealed that they all have a similar structure: a circular relief dominating a central depression half filled with debris, in the center of which is a little hummock. An irregular dusting of rocky fragments, gravels, and sands of various sizes covers the ground around the crater—the crater's influence zone. We can thus imagine that a series of adjacent craters would produce a stratification of ejecta layers—an actual impact stratigraphy for the moon.

The craters exhibit a general relationship between their *frequency* and their *diameter*. As a first approximation we can say that the number of craters increases in geometric progression as the diameter decreases. In other words, for every large crater there are innumerable small craters.

The study of the phenomena of craterization was first attempted by experimenting on small models in which substrates of various composition were bombarded with projectiles shot from canons at various speeds. Donald Gault of NASA's Ames Research Center was a pioneer in this type of spectacular but very tedious experimentation. He was able to establish a precise relationship between the diameter of a crater, the mass of the impacting projectile, and its speed. These relationships showed what kind of projectiles created the lunar craters. The importance of this will become clear later on.

The density of craterization rapidly became a chronologic tool using the following simple principle: if region A has a greater crater density than region B it is geologically older than B. This method of counting craters was subject to complications—notably, that one had to allow for a certain saturation in heavily bombarded areas—but these difficulties were surmounted, and the method can now be used to establish the relative chronology of any planet.

The radioactive chronology of the lunar rocks and of the various maria, or lunar seas, was used to quantify these observations. When crater density was plotted as a function of age (Figure 27), it became clear that the rate of bombardment had decreased exponentially over the last 4.5 billion years. That is why the lunar mountains are much more cratered, and therefore more chaotic-looking, than the maria. The very marked decrease between 4.5 billion years ago and about 3 billion years ago is undoubtedly one of the major findings of the lunar exploration. It shows that the phenomenon of planetary accretion, which culminated about 4.5 billion years ago, fell off very rapidly after that, probably because of a paucity of projectiles, since the

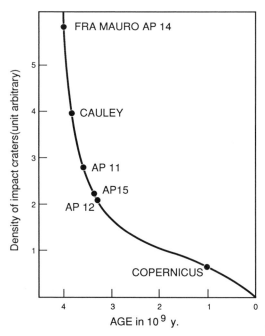

Figure 27 The curve shown on this graph is one of the most important results of the moon exploration. The density of impact craters decays strongly with age, as determined by radioactive clocks, which illustrates the history of accretion.

majority had been captured by planets and had taken part in their formation.

The primitive bombardment was very intense. The diameter-frequency relationship of the craters suggests that there were gigantic impacts at that time. What evidence is there of such impacts? The shape of the maria with their circular nested contours suggests an answer. The maria could have been created by gigantic impacts nearly 1,000 kilometers in diameter. The application of theoretical formulas leads to the conclusion that the projectiles that caused these depressions had diameters of 100 kilometers and that they excavated quantities of material on the order of 10^{24} grams, about the mass of the continental crust of North America. The amount of heat generated by such impacts would have melted the lunar interior to produce magmas that would have filled the depressions, thus forming the basaltic maria.

This theory was developed to explain observations made by the space missions, with no connection to the Soviet theory. George Wetherill (1976), who was then working at UCLA, connected the selenological conclusions and the Soviet theory of accretion. The impacts of the meteorites, whose size and frequency decreased between 4.5 billion years ago and the present, marked the end of the accretion phenomena, the "accretion tail," as Wetherill noted. In fact they support the Soviet theory: planets form through the addition of previously formed balls of different sizes, and this process is not instantaneous but takes millions of years.

If Western theorists had treated the Soviet model with disdain up to that time, Wetherill's theory immediately attracted their attention. Having adopted the Soviet theory, they began to improve upon it and make it more precise by using the powerful computers to which Schmidt had never had access. Computers were able to simulate experiments easily, reproducing the snowball fight under various conditions. Of course, these were not experiments in the usual sense of the term but simulations or scenarios calculated step by step, situation after situation, as modern computers permit us to do. The work was long, difficult, and tedious, but these scientists were able to refine the scenario imagined by the Soviets in several areas: the duration of the phenomenon, the number of planets, the thermal regime of the planetary bodies, and the accretion conditions.

Computer simulation showed that it is rather "easy" to go from dust to objects about a kilometer in diameter, and in the conditions of the primitive solar system this process would not have had to take more than a million years. On the other hand, to go from objects of this size to planets is a "difficult" operation that, when it succeeds, takes 50 to 100 million years. This makes sense if we remember that the larger the objects become, the fewer there are and the fewer chances they have to meet. When this stage is not reached the objects not only cease to grow, but as we shall see, they break up into increasingly smaller pieces.

Another accomplishment of the computer simulations was to specify the thermal consequences of the collisions. When two object collide with each other, they heat up. The heating can be calculated as a function of the characteristics of the projectile. Within the framework of progressive planetary accretion it was shown that the thermal energy transferred to the planets during formation by the impact of the projectile could cause an almost total melting, or at least set off surface volcanism. The hypotheses for the formation of the maria had thus tested positive as a result of computer simulations.

Numerical calculations made it possible to introduce an additional factor in the accretion process: gas. This is the gas that is often mentioned in connection with volatiles, but which we have overlooked until now. It appears that gas plays an essential role, especially at the beginning of the accretion process, by decreasing the speed of small fragments and cushioning collisions. Moreover, small particles that are moving about rub against the gas, slow down, and are heated. John Wood (1984) even thinks that these friction phenomena are the source of the formation of the small drops of cosmic magma that cool into chondrules, the characteristic globular structures in chondritic meteorites. Gas thus facilitates the process of dust accretion. Although problems remain, the model of progressive accretion has made it possible to relate many observations that had previously been unconnected.

Geologic Impacts

The extrapolation of the lunar results to the other planets is more problematical, since the density of projectiles clearly depends on the distance from the sun. But the Soviet theory comes to the rescue. Using the collision model, Wetherill was able to establish a relation-

ship between the bombardment curve of the moon, directly calibrated by the dating of lunar rocks, and the hypothetical curves of
the other planets. Even these calculations were tested, since Wetherill
was able to predict the density of craters that should be found on
Mercury several months before the Mariner 10 mission. In any case
the impact phenomena exist everywhere, from Mercury to the satellites of Saturn, from the moon to Mars's satellite Phobos. We have
encountered them constantly in our exploration of the solar system.

It is certain, therefore, that comparable collision phenomena acted
on the Earth at the dawn of its life. Our planet underwent continuous
bombardment by bodies a few meters to 100 kilometers across. Coming from all directions, these projectiles pocked the surface of our
planet, making it look like a battlefield after the continuous action of
heavy artillery. Every impact created a crater. The large bodies created basins the size of the Michigan basin; the smaller ones contented
themselves with blasting craters a kilometer in diameter; and the
smallest ones, which were much more numerous than the large ones,
made innumerable little scars of metric dimensions. Of course, each
impact crater was then subject to further bombardment, which
destroyed its edges and its shape. The immediate effect of this bomb-
ing raid was to turn the Earth's surface into a chaotic and dusty jum-
ble. The rock debris, breccias, and rubble piled up at the bottoms of
slopes so that the Earth's surface looked like a demolition site. This
apocalypse lasted almost 500 million years, but with rapidly decreasing intensity. If its traces are less visible today than they are on the
moon, it is because the Earth is a geologically active planet, and subsequent geologic phenomena have erased the traces of the landscapes
of its earliest history ("no vestige of a beginning . . ."), although a
whole series of ancient eroded craters as well as modern craters like
the majestic Meteor Crater in Arizona still exist. The geologic role of
the craters in the earliest periods is indisputable, but what happened
after that?

Has the number of meteorites falling on Earth during the course of
geologic time been sufficient to increase its mass significantly? In
other words, is the Earth growing constantly? The inventory of present-day meteor hits and the calibration of lunar craters suggest a negative response to what could have been a major discovery (the origin
of continental drift in the expansion of the Earth, as S. Warren Carey
of Australia believed). The Earth has gained only 10^{25} grams of extraterrestrial matter over 4 billion years—which is not so bad, since it

equals the mass of the continents. But it makes a negligible difference, nevertheless, in the total terrestrial volume, since the Earth's mass is actually 6.10^{27} grams. Let us look for more modest effects. If the impacts created lunar seas 3.2 billion years ago, why couldn't they have created comparable terrestrial structures?

For a long time geologists have noticed that Precambrian terrains are the only places that contain certain special large rocky formations called layered intrusions. These formations can be hundreds of kilometers wide. One of them, the Bushveld in South Africa, is 300 kilometers across. The rocks that form them contain large crystals, like granites, but their composition is very different from that of granites. It is a combination of rocks whose composition is sometimes like that of basalt and sometimes like that of peridotite (that is, mantle rock). They therefore originated from deep within the Earth. These simatic (silicon and magnesium) massifs have long been the object of particular interest because they contain considerable mineral resources. One of them, the Sudbury, furnishes Canada's supply of nickel and chromium; the Bushveld contains the largest platinum reserve in the world as well as chromium; others in the Soviet Union are just as rich in precious metals but are surrounded by great secrecy. How did such formations come to be? Why do they exist only in ancient terrains?

In a detailed study of the Sudbury, Franck and Dietz found indisputable traces of a gigantic 2.5 billion-year-old impact on its borders. They concluded that these formations, which no geological theory explained convincingly, might all be the result of gigantic impacts that acted like those that produced the maria. On Earth we observe only their interiors, since erosion has worn away their edges. This theory, which has been accepted for the Sudbury, has not been proved for the Bushveld (2 billion years old) or the Stillwater Complex in Montana (2.7 billion years old), and thus remains conjectural. Was the geological role of impacts, like that of the simatic massifs, confined to the ancient period before 2 billion years ago?

The late Luis Alvarez, a well-known Berkeley Nobel Prize winner in physics, and his son Walter, a geologist, discovered that at the exact boundary between the Cretaceous period and the Tertiary period in layers 65 million years old, there is a thin layer in about twenty places that is rich in metals associated with platinum, especially iridium (Alvarez et al., 1982). Terrestrial rocks are very poor in these elements, while meteorites are relatively rich in them. The

Alvarezes concluded that the Earth had been uniformly covered by a
layer of meteoritic dust 65 million years ago, no doubt as a result of
the impact of a giant meteor.

This would not be particularly important were it not for the fact
that the Cretaceous-Tertiary boundary also corresponds to a major
biological transition. At that time, in fact, the ammonites, the dino-
saurs, and more than a thousand marine species disappeared from
the face of the Earth. As we know, the sudden disappearance of species is one of the most difficult problems for paleontologists. The
Alvarezes, father and son, related these two phenomena and concluded that the fall of a meteorite was responsible for the extinction
of these species: dust in the air blocked sunlight and caused a generalized cooling of the climate. Not surprisingly, this assertion set off
a vigorous debate between the Alvarezes and their supporters, and
some of the most eminent paleontologists. Without going into that
debate, I would point out that these ideas are revivals of those proposed by Georges Cuvier a hundred and fifty years ago.

We are forced to admit that the chronological species extinction
diagram made by the paleontologists themselves does show extremely well-defined sharp peaks (see Figure 28). Each peak, each

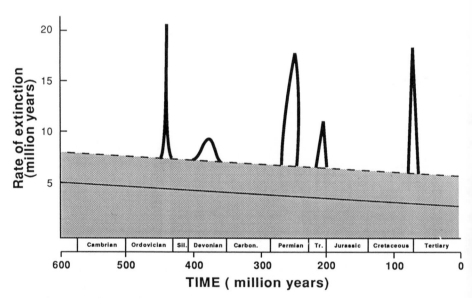

Figure 28 This graph shows the rate of the extinction of species through
geological time.

episode, corresponds to massive extinctions, to real biological catastrophes. What does this mean? It is in no sense a refutation of the theory of evolution but only a recognition that one of the factors that has played a role in natural selection is extraterrestrial in origin. The species that survived were those best able to adapt to unusual events, not those best adapted to normal events.

The Origin of Meteorites

In Chapter 4 we saw how a few stones that fell from the sky, a few meteorites, have helped us considerably in our study of the processes of generation. Their fall really appeared to be providential—a gift from the heavens! Carbonaceous meteorites, chondrites, and differentiated meteorites all provide evidence of the formation process of planetary material. It is as if Nature had taken a sample for us at each step of the planetary formation process, had kept it intact somewhere for 4.5 billion years, and had finally sent it to us from the sky so that we could study it in our laboratories.

The existence of these primitive witnesses to the first instants of the solar system, which have the good taste to fall onto our planet periodically, seems to have a touch of the supernatural. We cannot avoid asking: Where do they come from? How were they formed?

Through direct ballistic observation of some of their trajectories as they fell to Earth, we know that most of the meteorites come from a region of the solar system between Mars and Jupiter where there is no planet (see Figure 29). Called the *asteroid belt,* this region is populated by myriad objects whose size ranges from that of a stone to that of a planetary body. The largest of the asteroids, Ceres, is 1,000 kilometers in diameter.

All the rocky objects in this belt, circulating in orbits that are very close together, collide with each other and overtake each other, and since they are moving very fast, these shocks cause them to break apart. Thus, over time, the asteroids become more fragmented and the number of small pieces increases. According to the ballistic evidence, meteorites are pieces of asteroids that have left their usual trajectory and fallen to Earth. This hypothesis was confirmed by telescopic observation, which showed that the optical properties of the asteroids are similar to those of the meteorites that have been analyzed in laboratories.

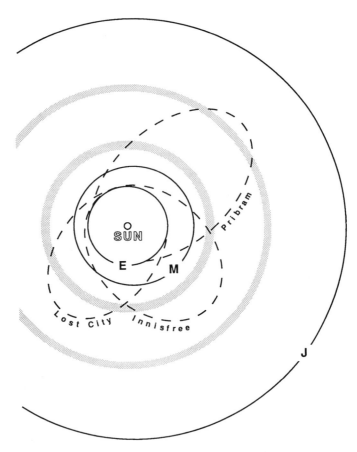

Figure 29 Three trajectories of meteorites that fell to Earth, as reconstructed by astronomical observations and calculations. The orbits of Earth (E), Mars (M), and Jupiter (J) are shown. Between the two gray circles is the zone of the asteroid belt.

Meteorites are thus fragments of larger-size objects that have been broken up by collisions in the asteroid belt. But where did those objects themselves come from? Could they be pieces of Orchelt, the tenth planet imagined by Plato to satisfy the omnipotent presence of the number ten, which Bode's law on the spacing of the planets situated precisely between Mars and Jupiter at the exact location of the asteroid belt?

As exciting as it is, this hypothesis does not fit the observations.

The chemical and mineralogical diversity of the meteorites makes it almost impossible that they could have come from the same planetary object. It has been shown that at least five objects—five protoplanets—would be necessary to account for this diversity. Besides, when we try to determine the conditions of temperature and pressure in which the meteorite minerals were formed, we realize that, aside from those formed in the zone of intense shocks, they are not minerals that are formed at high pressure, they are those that are formed only at rather low pressure.

The absence of high-pressure minerals in meteorites (except in shock zones) suggests that the parent bodies were not very large. A more precise determination of their size can be obtained by studying their cooling rates. Iron phases in meteorites are in fact iron-nickel alloys. Actually two different alloys exist, one rich in nickel and one poor in nickel. When iron meteorites cooled, the two types of iron-nickel alloys recorded this thermal evolution and a reaction zone developed at their borders. Jo Goldstein of Lehigh University calculated the cooling rates by studying the nickel variation in this reaction zone. With John Wood of Harvard she extended this method to chondrites by using the pieces of iron they contain. More recently, a different method based on the tracks of plutonium fission in crystals was used by Paul Pellas of the Paris Museum to check Wood and Goldstein's figures. Despite differences in absolute numbers, all these studies indicate a rapid rate of cooling and therefore relatively small bodies; the meteorites' parent bodies were thus smaller than the moon and all together their volume is much smaller than that of Mars. It was not a large dislocated planet that engendered the asteroid belt and the meteorites but more probably several small, distinct planetary bodies.

These primitive objects were many and varied. Some were agglomerations of primitive matter whose fragmentation produced chondrites. Others were microplanets, with cores in their centers and mantles and volcanic activity on their surfaces, whose fragmentation produced differentiated meteorites. The fragmentations must have been extremely sharp and deep, breaking the objects almost through to their centers, their very hearts, since in the form of meteorites they now show us their most intimate composition. The question of chronology arises. Exactly when did this free-for-all take place? Did it occur all at once, or were there several episodes?

The Exposure Ages of Meteorites

A flux of very high-energy particles is continuously passing through cosmic space. The most numerous and energetic of these are the protons, whose source—although unknown at present—is exterior to the solar system. This radiation is called galactic cosmic radiation (GCR) to distinguish it from the radiation of less energetic particles emitted by the sun. The high-energy protons penetrate matter and set off nuclear reactions that transform certain elements or isotopes into other elements or isotopes. This penetration is not very deep, however, and does not exceed a few tens of meters.

A free rock in the cosmos is thus subject to this cosmic flux, whereas a rock in the interior of a planetary body is protected from it. When we detect highly abnormal isotopic compositions in a rock, it suggests that it was isolated in the cosmos and that its surface was thus subject to galactic cosmic rays. The detection of abnormal isotopic compositions is one of the best criteria for determining whether a rock found on Earth is a meteorite. For example, pieces of iron meteorites have quite abnormal isotopic compositions of calcium, magnesium, and potassium, while chondrites have an abnormal isotopic composition of rare gases. This abnormal isotopic composition comes from a well-known nuclear reaction of iron or silicate minerals. As we would expect, the longer the irradiation, the more "artificial" isotopes are formed. The number of isotopic anomalies is a function of how long the irradiation lasted. This is called the *exposure age* to reflect the fragmentation age of the meteorite, the point at which it became a free rock and thus subject to irradiation.

The results of this dating are extremely interesting. For silicate meteorites of the chondritic or achondritic type, the exposure ages are 20 to 150 million years. For iron meteorites, they vary from 50 million to 2 billion years. We should remember that all these meteorites have the same *formation age,* 4.5 billion years, which is much earlier. The exposure ages show that the fragmentations took place long after the formation period. The variability of the exposure ages and their maximum length indicate clearly that the collisions and fragmentations have been taking place uninterruptedly for several billion years and are ongoing and usual events in the rough and tumble careers of the asteroids. The iron meteorites resist this cosmic pinball game best because they are the hardest, so they have older exposure ages. Silicate meteorites, and in particular the more fragile and crumbly chondrites, break up into smaller and smaller pieces, so their exposure ages are younger.

Now we can try to understand why the asteroid belt contains innumerable different objects instead of a single planet. Let us try to reconstruct what could have happened about 4.5 billion years ago when everywhere in the solar system planetary bodies were agglomerating. In the asteroid belt many planetary bodies started to agglomerate. Some remained modest in size and therefore in the state of aggregates, of primitive matter. These are the original source of the chondrites. Others reached a larger size, accumulating sufficient energy to raise their internal temperature enough to induce fusion, leading to the formation of a central core and setting off surface volcanism. None of these bodies, however, had enough gravitational energy to attract others and form a single planet.

The accretion process slacked off and then reversed. At a certain point, instead of adhering together after a collision, the different bodies broke up and fragmented, eventually creating the swarm of disparate asteroids we observe today. The existence of these shocks can be seen in the meteorites themselves: the internal structures of many of them contain sharp, angular fragments of a kind of breccia.

This touches on the fundamental phenomenon already suggested: the existence of a continual state of transition, a delicate equilibrium between accretion and fragmentation. The first is the constructive encounter of objects in space, the second their destructive collision. Why did the equilibrium tip toward accretion for all the planetary objects except one? Perhaps the proximity of Jupiter, which had already been formed, disturbed the process?

In a somewhat analogous situation, series of satellites exist around the giant planets, as well as rings made of myriad small rocky objects. Like the asteroids, the rings suggest that in some places fragmentation replaced accretion. The rings are the equivalent of the relationship between the asteroids and the sun on the scale of a planet the size of Jupiter or Saturn.

The play of collisions thus reveals a great variety of situations to us, but the game is not yet over.

Interplanetary Voyages

Antarctica is a great frozen desert; the snow that falls there is very pure. If a rock is found sitting on this white carpet, there can be no doubt that it fell from the sky. When we realized that the Antarctic is a natural collector of meteorites, we decided to explore it in order to bring back clean, well-preserved specimens gathered by the glacial valleys that, like moraines, contain boulders and unsorted rocks. Surprising results came out of this new kind of "space" exploration.

In 1975 an American expedition discovered a meteorite in the Antarctic whose chemical characteristics resembled neither the usual chondrites nor the achondrites. Analyzing the isotopic compositions of the rare gases and xenon-129 it contained, Robert Pepin of the University of Minnesota realized that they were similar to those that had just been measured on Mars by the Viking mission. The measurement of the isotopic composition of nitrogen, which is one of the characteristic signatures of the Martian atmosphere—the light isotope escaped while Mars was in its hot phase—fully confirmed the proposed hypothesis.

Then it was recalled that in the collection of achondritic meteorites there were two, called Nakla and Shergotty, with unusual characteristics: they were 1.4 billion years old while all the others were 4.55 billion years old. In studying both, Richard Becker and Pepin (1984) discovered that they were identical to the meteorite found in the Antarctic and had the "Martian" chemical and isotopic signatures. So at a certain time the collision of a large meteorite must have torn a piece from the Martian crust. Its expulsion speed (5 kilometers per second) was greater than the escape speed of Mars, and the rocky chunk was thrown into space in a more or less complex trajectory on a voyage that lasted several million years (determined by the exposure age) until it fell to Earth near the city of Nakla in Egypt.

But the surprises didn't end there. More recently a real lunar rock was found in Antarctica. The comparison was easier to make in this case, since we already had a good collection of lunar rocks in Houston. As analysis demonstrated, without doubt this was another natural space rocket. Thus, it must be accepted that the impact of a giant meteorite on Mars or the moon was sufficient to blast rocks out of the ground and hurl them into the air so violently that they escaped their planet's field of gravity and after several million years of wandering fell to Earth. Men thought they had invented interplanetary voyages, but Nature had preceded them by many million years! In the past we have spoken of Newtonian clockwork; now we must speak of the "Newtonian planetary billiard game."

The existence of impacts gigantic and powerful enough to excavate pieces of a planet gave birth to daring hypotheses. Some thought the moon had been snatched from the Earth by a huge meteorite that hit the planet after its core had been differentiated. That would explain the low iron content of the moon. The shock would have caused the volatile elements to vaporize, which would account for their low abundance in lunar rocks. The moon would thus be a daughter of Earth, but by forced extraction. This hypothesis, defended by Henrich Wänke of Mainz, Germany, and Ted Ringwood of the Australian National University, is interesting and explains many of the chemical observations we have mentioned. Of course, it is disputed by others.

Unlike all the other planets, Venus turns backwards and very slowly. The reason for such anomalous behavior is still a mystery. Some think the explanation is that Venus, which initially turned in the "right" direction, was hit by a cosmic body that turned its axis in the opposite direction like that of a spinning top. Having become retrograde the rotation would slow down gradually and would become "normal" again in a few billion years.

Clearly, the effects of impacts catalyze the scientific imagination!

Comparative History of the Planets

Without chronology there can be no history. A chronology based on counting impact craters and calibrated by dating lunar rocks has made it possible to attribute absolute ages to the various terrains and to make "geologic" maps of the principal planets. These maps summarize the history of the planets (see Figure 30). What do they tell us?

- Geologic activity on the moon, activity whose origin is internal, ended about 3 billion years ago, when the seas were invaded by huge basaltic lava flows.
- Mars's activity seems to have culminated about 2 billion years ago with the birth of great shield volcanoes such as Tharsis and Olympus Mons. The geology of its recent periods seems to have been limited to phenomena of superficial fluvial or aeolian origin, that is, to an external geology.
- Mercury's activity seems to be much more primitive yet, and although calibrating the counts of its craters presents a few

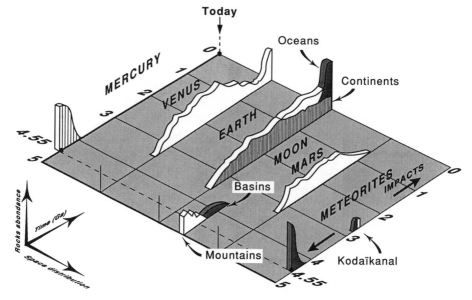

Figure 30 Histograms of the ages of the surface rocks of different planets as determined by the density of craters for Mercury, Mars, and Venus (data for Venus are the most speculative of all) and for the Earth, moon, and meteorites by radioactive clocks.

problems, it is estimated that its geologic activity ended 4 billion years ago while the moon was still in its active phase.
- Jupiter's satellite Io remains active today. We observed a volcanic eruption there during the Voyager mission.
- Venus also seems to be a geologically living planet. Recently the Arecibo telescope is thought to have detected indications of an explosive volcanic eruption there.

Meteorites, on the other hand, whether they are chondrites or differentiated meteorites, began their lives sometime prior to 4.4 billion years ago and have traveled through time without any notable alteration.

Thus, little by little, it became clear that the various planets and their satellites "jelled" at different stages of their evolution. And it is tempting to say that the history of the Earth can be reconstructed by putting the various scenarios of planetary development together one after another. Considering the different classes of meteorites, one might say that *planetary phylogeny recapitulates terrestrial ontogeny.* This view is, however, only a hasty simplification, evidence of our impatience to understand our origins in order to justify our extraterrestrial enterprises but also to cling to the simplest possible explanation, which is always at the back of our minds.

In fact each planet's history is different. The fact that the planetary evolutions cover different periods and give us sequential records of the activity of the solar system is useful only for comparative analogical reasoning; it does not provide strict parallels.

While it is true that each planet gives us important information applicable to Earth, a planet cannot be treated as a simple projection or copy, because the giant clockwork regulated by the laws of universal gravitation does not produce identical objects of the same composition and behavior, differing only in size. The histories of the planets are different because their internal composition—their chemistry—is different. The Soviet snowball-fight theory explained the diversity in the size of the objects formed. Now we must explain the diversity in their chemical composition, in their energy sources, and in their resulting geological behavior.

Comparative Structures of the Planets

By applying the principles of geometry and celestial mechanics, spatial exploration has precisely determined the radii of the planets, their density, and the way in which the mass is distributed in their interiors, a property that is translated in the parameter called *moment of inertia.* As we have seen, scientists in the nineteenth century were able to predict an internal structure for the Earth that was brilliantly confirmed and refined by seismology. From then on it became legitimate to apply the same methods to the other planets and satellites to try to determine their internal structures (Figure 31).

The planets are supposed to consist of successive layers of decreasing density from center to surface. The assumed materials are, for the solids: iron, iron sulfide, and silicates; for the gases and atmospheres, compounds whose presence was detected during spatial exploration: water, nitrogen, carbon dioxide, ammonia, methane, hydrogen, and helium. The results of calculations by successive approximation exhibit a fundamental fact: there are no two identical planetary structures, no two planets whose various layers have the same composi-

a

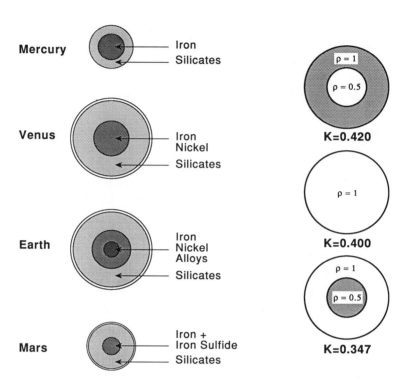

Figure 31 a) The internal structures of the inner planets determined by their bulk density and moment of inertia, assuming that all these planets are made of iron, iron sulfide, and silicates. b) Three planets with different structures (K indicates the coefficient of the moment of inertia).

tion and dimensions. There are not even two similar planets; each has its own characteristics. Mercury is different from Venus, Venus from the Earth and from Mars, Jupiter from Saturn, Io from Ganymede, the telluric planets from the giant planets, and the moon from the satellites of Jupiter and from Mercury.

The other dominant characteristic that emerges from this catalogue of the internal structures of the planets is a zonality in relationship to the sun. Mercury, which is near the sun, is dense, rich in iron, and without an atmosphere. Venus, the Earth, and Mars all have a core, a mantle, and a rather weak gaseous envelope, with few or no satellites. The giant planets, whose field of gravity retains even hydrogen and helium, have gigantic atmospheres and a horde of satellites, which are arranged around their "home" planet in order of decreasing density.

The diversity and variety that are so apparent in the internal structures of the planets are just as clear in the chemical composition of their atmospheres. The atmospheres of Venus and Mars are rich in carbon dioxide; that of the Earth is dominated by nitrogen and oxygen, of Titan by methane and ammonia, and of Jupiter by hydrogen and helium. The Earth's water-cloud meteorology corresponds to an acid-cloud meteorology on Venus, "oil" on Titan, and helium on Jupiter. There are as many different regimes as there are planets.

There is also a diversity in volcanism, which is, after craterization, the most widespread phenomenon in the solar system. The silicate lavas of Mars and the Earth correspond to the sulfured volcanism of Io and the "fountain-volcanism" of Ganymede and Enceladus. The volcanism of the Earth and Mars as a product of the radioactive disintegration of uranium, thorium, and potassium corresponds to the moon's volcanism, which is produced by heat from impacts, and Io's volcanism, which is produced by the Jovian tides. Thus is a variety of causes hidden behind the uniformity of appearances. The existence of variable magnetic fields, notably for Mercury, the Earth, and Jupiter (those of Venus, Mars, and the Jovian satellites are negligible), is also evidence of a great diversity of conditions among the planetary bodies.

Within this group the Earth appears to be a "rich" planet, first because it is the only one to support life, but also because it possesses both a crust-mantle and a crust-atmosphere differentiation, a magnetic field, and external and internal geologic activity that has lasted from its beginnings to the present day. Some people may draw philosophical conclusions from this; others will see in it yet another difficulty in our attempts to understand its beginnings.

Systematic observation of the planets shows that from a common origin they developed a great diversity, in both structure and geologic history. What characterizes each planet and is no doubt at the source of the diversity is chemical composition. A heliocentric chemical zonation gives each planet its own chemical character, its specific composition. As Urey has noted, understanding the formation of the solar system is not only a problem of mechanics but also of chemistry.

Mechanics governs the complex motions of the planetary objects, whose clockwork seems the essence of classical mechanics. Chemistry introduces variety, diversity, and fantasy, but not anarchy, because, as we have pointed out, heliocentrism remains the rule, and the chemical composition of the planets is not a matter of chance. To understand the formation of the solar system we must therefore reconcile Newton and Mendeleyev. To do this, in the framework of celestial mechanics, is to make room for chemical diversity and its multitude of consequences for the specific character of each planet.

The collision theory developed by the Soviet school has made it possible to explain the planetary movements, the accretion of planets and meteorites, the phenomena of impacts, the fall of meteorites, and the cosmic "billiard game," all with the same scenario. It is the triumph of Newton, with Schmidt as his intermediary. But this theory stumbles over one difficulty: it does not explain the chemical variety, the planetary diversity.

The Hot Nebula

Under the influence of Urey and his disciples and students, a *unified theory* was developed that accounts for all the characteristics of the planets as well as those of the meteorites discovered during the course of the "space" decade. It is the *theory of the condensation of the protosolar nebula* (Cameron, 1963; Grossman and Larimer, 1974; Anders, 1971).

The point of departure is the protosolar disk, turning on its axis and inflated at its center as Kant and Laplace had supposed it to be. The nebula consists of hot gas. Its chemical composition is almost the same as that of the sun today, since the sun alone contains 99 percent of the mass of the solar system. Let us follow the "natural" evolution of this system. The central ball heats up more and more as it contracts so that its temperature soon exceeds that of the surrounding disk. A sharp thermal gradient between the center and the edges of the disk is established. The hot disk gives off light. It therefore loses energy and cools around the edges. When the gas reaches its condensation temperature, solid particles form, following the condensation sequence described in Chapter 4. Now we no longer have a gas, but a mixture of gas and dust. When the density of the dust becomes high enough, it begins to agglomerate into solid objects the size of marbles.

These are the familiar planetesimals, whose evolution through accretion has been outlined here.

The cooling leading to condensation into solid grains is not uniform for the whole disk. Regions on the frontiers between the disk and interstellar space cool more rapidly than those near the newborn star that is the proto-sun. Hence, at any given moment the temperatures of the various parts of the proto-solar disk are different. Now we know that at each temperature below a certain threshold a mineral of a given chemical composition condenses. When the temperature of the gas falls, the series of solids described in the condensation sequence is deposited: refractory oxides rich in aluminum and titanium as found in the Allende meteorite, then metallic iron with nickel, then silicates rich in magnesium, and so on. Thus, near the sun, at the distance where Mercury will be formed, the temperature is 1,200°K, while at the distance of the Earth (one astronomic unit), it is only 1,000°K. At Mercury there are solid grains consisting of iron, aluminum oxide, and a few silicates, while in the terrestrial atmosphere bits of iron are found but also a lot of olivine and serpentine, which are silicates rich in water. Suppose that the solid grains then agglomerated to form a planet: Mercury would be rich in iron and contain a little aluminum oxide and silicates, while the Earth would contain a large proportion of silicates. We would therefore have two planets of different chemical compositions. Between the two, Venus, which is similar to the Earth, would have little water. This is roughly the scenario of planetogenesis that is proposed (Lewis, 1973) (see Figure 32).

To account for the properties of the giant planets and their satellites, John Lewis has calculated the condensation sequence for the lower temperatures. He has shown that ice condenses from a gas of solar composition at 0°C, followed at lower temperatures by carbon dioxide and then by methane and ammonia. All the *observed* components of the giant planets and their satellites have been accounted for. The chemical zonation of the solar system can thus be explained by assuming that at a given moment the sun formed and that like all stars in the process of formation it emitted a violent wind of particles that blew away all the gases present in the adjacent solar system. Only the solid grains that had been accreted into planetesimals remained. Since this episode, called T-Tauri (the name of young stars of the same name), took place when the temperature in the region

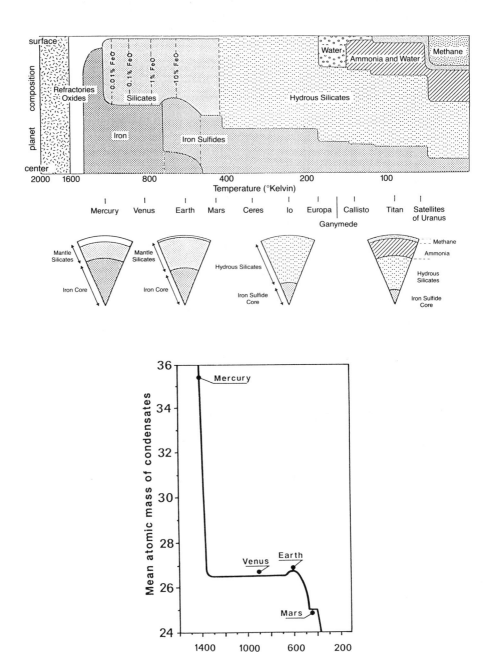

Figure 32 John Lewis's (1973) scenario for the formation of different planets by equilibrium, condensation, and accretion. The size of the planets has been assumed to be the same, and their density has been plotted as a function of the temperature of condensation.

where Mercury would form was about 1,100°K, this planet contains only iron and a few silicates. The temperature in what would become Venus's orbit was then about 500°K, so the proportion of silicates is much greater there. Formed in a slightly colder zone, the Earth accreted silicates containing water. For Mars the proportion of iron is much lower compared to the silicates and sulfides that were condensed.

Only in the region of what would become the giant planets was the condensation temperature for water reached. Everything in this region was condensing; chemically the giant planets are analogous to the sun, but they are not large enough to become stars. The heating created by their contraction was not great enough to set off nuclear reactions. It was enough, however, to create a thermal gradient and zonation in the satellites that would condense around them. The phenomenon of zonal condensation in relation to a center would therefore be repeated: Io, near Jupiter, contains a lot of rock but little water, while Europa, Ganymede, and Callisto contain a lot of water, little rock, and perhaps a little ammonia and methane mixed with ice. We understand why Saturn's closest satellites are covered with ice while Titan contains much more methane than ammonia. But how can we explain the presence of an atmosphere on Venus, Earth, and Mars? We can explain it by assuming that gases trapped in solids and agglomerated with them will degass during secondary phenomena such as volcanism, which, we know, always gives off large amounts of gas. This scenario, proposed by John Lewis, assumes homogeneous, uniform accretion of the materials in a given nebular region.

Unshakable proponents of heterogeneous accretion like Sydney Clark and Karl Turekian of Yale propose an identical scenario, but with the accretion beginning as condensation. Thus, as soon as iron has condensed, it accretes into a core. Condensed silicates accrete around the iron cores producing planetary mantles, and finally volatiles condensed to ice are the source of atmospheres. The wind of the proto-solar T-Tauri phase stopped the condensation-accretion phenomenon for the internal planets and blew most of their atmospheres into space.

This scenario offers various advantages. It explains the moon's iron deficit well. Since the moon condensed outside of the proto-solar disk in a much colder area, its iron content would be lower, just as that of the Earth is lower than that of Mercury. On the other hand, it cannot explain the atmospheres retained by the telluric planets, for which it is difficult to conceive a primary origin, as we shall see. It does not explain the abundance of chondrites, formed from a mixture of all phases of iron and silicates, which seems to support homogeneous accretion, or the presence of feldspars, which are abundant on all the solid planetary bodies but are not included in the direct condensation sequence and can be formed only by secondary reactions.

It is as if the truth contains parts of both theories.

In any event, around 1978 the whole scientific community was persuaded that it was approaching its goal and that in the mixed condensation-accretion theory it had a scenario that explained the astronomical observations and the mechanical and chemical constraints of the solar system. The decade of planetary exploration was ending in apotheosis. The quest for the holy grail, the unified planetary theory, seemed on the point of attainment.

7

The Cosmic Palimpsest

None of the planets we have explored is made of pure gold, nor is Uranus made of uranium. All the inner planets are composed of the same elements: oxygen, silicon, magnesium, and iron. All the giant planets consist of hydrogen, helium, oxygen, carbon, and nitrogen (a composition similar to that of the stars). The cosmos contains ninety-two different chemical elements but only ten or twelve were used to build its structure. Why?

The answer to this apparent paradox is that the elements are not equally abundant in the cosmos. The objects in the cosmos are made of a few chemical elements because these are the most abundant. If uranium or gold were as abundant as silicon or iron, perhaps we would have planets made of uranium oxide or pure gold.

Let us examine the abundances of the elements in the universe (see Figure 33). When atomic mass increases, abundance generally decreases and does so very quickly. Hydrogen, the lightest of the elements, is by far the most abundant; helium is next but is less abundant by a factor of ten, followed by carbon, oxygen, and nitrogen. The heavy elements are very rare (10 to 100 billion times less abundant than hydrogen).

If we look at it closely, this decrease is not uniform. The curve has a fine structure with peaks and hollows, or zigzags, superimposed on it. One peak, among others, around the element iron upsets the general decay pattern of the abundance curve. Understanding this abundance curve is a major goal of science, because to do so is to under-

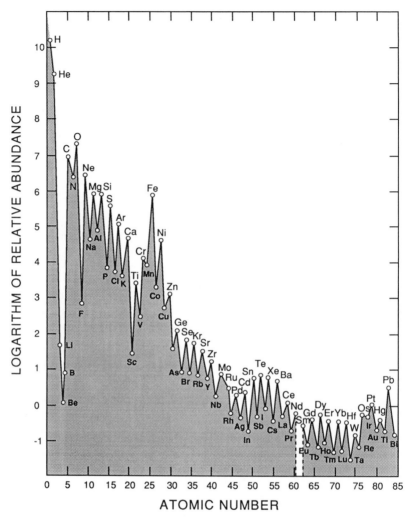

Figure 33 Abundance curve of the chemical elements in the cosmos. The scale is logarithmic. Note the lower abundances of lithium (Li), beryllium (Be), and boron (B) as well as the existence of a peak in iron (Fe). Note also the large differences in abundance between elements like carbon (C) and oxygen (O) compared with elements like the rare earths or bismuth (Bi). They are separated by seven orders of magnitude.

stand the ultimate causality that presided over the architecture of the universe.

The Origin of the Chemical Elements

At the end of the nineteenth century when scientists were trying to classify the elements according to their specific mass, Paul Prout, a French chemist, proposed a hypothesis considered very daring at the time. According to this hypothesis, all the elements resulted from the more or less complex combination of a single element, the lightest one of all, hydrogen: one element consisted of three hydrogens, another of four, another of five, and so on. The specific weights of the other elements must therefore be a multiple of that of hydrogen. The idea did not immediately take hold. It took more than fifty years and many detours to discover that Prout's hypothesis was essentially correct. And the confirmation came not from the chemists but from a group of scientists previously far removed from chemistry: astronomers or, more exactly, astrophysicists.

Actually, the elements are created in the stars. Gigantic nuclear reactors in which temperatures are measured in millions or billions of degrees, the stars are also immense chemical factories in which the various elements are concocted according to a complex alchemy. This discovery and its demonstration is an achievement of modern science that dates only to the last thirty years and was worth the Nobel Prize in physics for Hans Bethe in 1967 and William Fowler in 1983. To understand its development we must go back to the beginning of the century and the great adventure that started with the discovery of radioactivity, which was to lead, step by step, to the understanding of the intimate structure of the atom and of its stability.

The heart of the atom is the nucleus. Although it takes up very little space (if an atom had a radius of 100 kilometers, the nucleus would be a sphere only 1 meter across), it contains almost all the mass and therefore the energy of the atom (at this level mass and energy are equivalent and related by Einstein's famous equation $E = mc^2$). The transformation of one atom into another or the creation of one atom from other atoms, the dream of the alchemists of long ago, can only be effected by transforming the nucleus, the heart of the atom. (Nucleosynthesis = synthesis of the nuclei = synthesis of elements. In effect, once the nucleus is formed, the atoms always attempt to capture the peripheral electrons necessary to make it electrically neutral.) What modern physics has taught us is that nuclear transformations are possible. They can be spontaneous, transforming an unstable nucleus into one or several stable nuclei; these are then called *radioactivity.* They can also be set off by the collision of two unlike nuclei that break apart and recombine or fuse to give birth to new nuclei and therefore, new elements; these are called *nuclear reactions.* Nuclear transformations put considerable energy into play. To produce them in the laboratory we need giant machines like cyclotrons or particle accelerators. We can make use of this energy in terrifying atomic bombs or in more peaceful nuclear reactors. It is produced naturally in the stars, which are gigantic nuclear reactors.

The necessary link between high energy and nuclear transmutations explains both the failure of the alchemists and the success of the astrophysicists. The alchemists thought they could transform lead into gold by using the energy of a candle or a coal stove. But the energy put into play in their experiments was a few electron volts, just enough to modify the orbital electrons of the atom and transform the atom into ions or catalyze chemical reactions. To change the nuclei of atoms and thus attain their fantastic dream they would have had to produce energies a million times greater—in the range of Mev (one million electron volts) or even Bev (one billion electron volts). Alchemy is not chemistry. Chemistry can only combine already existing elements with each other. To make or produce them from scratch requires high energies like those used in nuclear physics.

Astronomers had estimated the amount of energy necessary to keep those powerful cosmic lighthouses, the stars, and the star that is closest to us, the sun, burning. They were looking for an energy source that could fuel these objects for billions of years. As soon as it became known that enormous energies are given off by nuclear reactions, astronomers thought about relating those reactions to stellar activity. This relationship did not escape the astronomer Arthur Eddington, a Cambridge colleague of the physicist Ernest Rutherford, the pioneer of the new nuclear physics. As soon as the two tried to quantify this simple idea, however, enormous difficulties appeared. The energetic barrier, born of the repulsion among nuclei (positively charged nuclei repulse each other) could not be overcome at the temperatures existing in the stars. In other words, the stars were not hot enough to set off nuclear reactions. Eddington yielded to the calculations of the nuclear physicists but continued to insist that the stars' energy source could only be nuclear.

It was only later, with the progress of nuclear physics, that these obstacles were overcome. George Gamow discovered the tunneling effect, in which a small number of particles obeying the laws of quantum mechanics could overcome apparently insurmountable energy barriers. The English astronomer Atkinson and the Austrian physicist Houtermans put this discovery to use to show that the small number of particles passing through the energetic tunnel were sufficient to ignite the nuclear fire. Independently, a few years later Hans Bethe and Carl von Weiszäcker of Germany consolidated this initial discovery by calculating a scenario of nuclear reactions that explained how the sun and certain simple stars could synthesize their energy.

The idea that the stars are the seat of nuclear reactions and thus fabricate new elements was confirmed directly when technetium was found to exist in certain stars. This element is unstable. All of its isotopes are radioactive with very short half lives (geologically speaking, which is 10^5 to 10^6 years). If technetium exists in a star, it must have been fabricated recently. This spectroscopic discovery totally validated Eddington's ideas and Atkinson and Houtermans's deductions.

Once the link between nuclear reactions and energy in the stars was established, it seemed likely that an explanation of the synthesis of the elements in the stars would soon follow. In fact, however, we had to wait nearly twenty years for this explanation to take shape, and then it was under the decisive direction of Fred Hoyle of England, and of William Fowler, Al Cameron, and Edwin Salpeter of the United States.

Without going into all the details of this rather complex chapter of astrophysics, which established the close relationship between the physics of the stars and nuclear physics, let us broadly trace the major steps.

Basic Elements in Nucleosynthesis

For contemporary physicists, the nucleus consists of hadrons and baryons, quarks and gluons with quite complex characteristics that range from "color" to "charm." For us, at this very elementary level, it is enough to know that the nucleus consists of two particles, the proton and the neutron. These two particles have almost identical masses but different electrical charges: the neutron is neutral, the proton positive. This difference explains the transformations the two particles undergo. A neutron can disintegrate spontaneously into a proton and an electron (with a negative charge) plus an antineutrino:

$$n^o \rightarrow p+ \; + \; e- \; + \; \bar{v} \quad \text{Example: } {}^{87}Rb \rightarrow \bar{e} + {}^{87}Sr + \bar{v}.$$

On the other hand, a proton can capture an electron and transform itself into a neutron plus a neutrino:

$$p+ \; + \; e \; - \; \rightarrow n^o + v \quad \text{Example: } {}^{40}K + \bar{e} \rightarrow {}^{40}Ar + v.$$

These two transformations play a fundamental role in nuclear physics and are the basis of two essential radioactivities, one called *beta* and the other *electron capture.* We can illustrate them simply by using a diagram in which the number of neutrons is shown as a function of the number of protons (see Figure 34).

Are the protons and neutrons in the nucleus present in random proportions and infinite variety? Definitely not, and we will soon see why. Since two protons have the same electric polarity, they repel each other. Nuclei that consist of several protons are therefore intrinsically unstable. The presence of the neutrons, a sort of "isolating link," stabilizes the structure. The role of the neutrons is such that they increase faster than the protons. While the ratio of neutrons to protons is close to unity for simple and therefore light nuclei, it is much greater than one for heavy nuclei. In Figure 35, which shows the number of protons as a function of the number of neutrons, the zone of naturally stable nuclei clearly departs from the line of equality between protons and neutrons as the number of protons increases.

We have spoken of stability. In effect, when a nucleus does not contain a ratio of neutrons to protons suitable for a given mass, it is unstable and transforms itself spontaneously, always tending toward stability. This phenomenon of spontaneous transmutation is called *radioactivity.* Its rhythm is unchangeable and independent of all physical and chemical factors. It is produced in three ways set out by Rutherford at the beginning of the century: radioactivity beta +, radioactivity beta −, and finally the explusion of an atom of 4He called alpha-rays. All this simply translates the existence of the complex nuclear interaction called weak and strong forces, whose nature

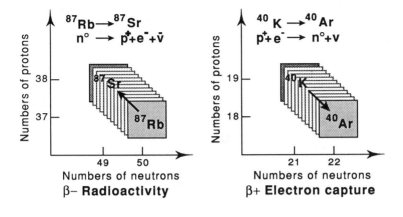

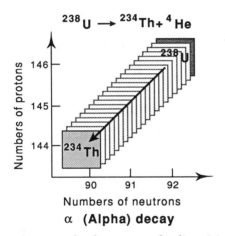

Figure 34 The three types of radioactivity.

we are now beginning to understand. These forces determine the complex logic of nuclear assemblages.

Assemblages of atoms can be grouped in various ways. The most useful groupings are based on the number of protons. All nuclei that have the same number of protons have the same positive charge (and the same number of peripheral electrons in their atom); they thus define the same chemical element. But sometimes nuclei with the same number of protons have different numbers of neutrons (in other words, the laws of nuclear stability are not too strict and do not

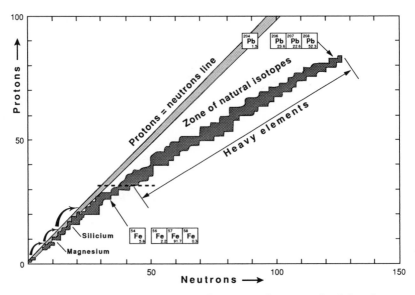

Figure 35 Protons-Neutrons diagram illustrating the zone of stability for natural nucleons. For the elements preceding iron (Fe), the number of neutrons equals the number of protons. For heavier elements neutrons are more abundant than protons.

dictate a neutron-proton ratio but admit some small variation). Although they characterize the same chemical element these atoms have a different mass. Atoms of the same element with different masses are called *isotopes* of that element. In nature each element is composed of a mixture of several isotopes.

Of course, each element has only a few stable isotopes. Some have only one; others, up to ten. These variations reflect the laws of nuclear stability, which involve parity in the number of protons, neutrons, nucleons, and so on—we don't need to know them at this stage. Outside of the stable range, an excess of neutrons contributes to instability, that is, to radioactivity. Most of the radioactive isotopes of an element are artificial. Those that could have been created naturally have decayed over the course of geologic time. The only ones that remain are those that have very long decay periods, so that a portion of them has survived the passage of time.

Not all the isotopes of an element are equally abundant. Natural oxygen is a mixture of stable isotopes of mass 16, 17, and 18 (written

^{16}O, ^{17}O, ^{18}O), whose relative abundances are very different. Sixteen grams of oxygen correspond to 6,231,023 atoms of oxygen. Of this number,

6.21479 × 10^{23} atoms have a mass of 16
2.4297 × 10^{20} have a mass of 17
1.2771 × 10^{21} have a mass of 18,

corresponding to the respective proportions of 99,756; 39; and 205 for 100 atoms. Rubidium, an example of the heavier elements, has two natural isotopes of mass 85 and 87 (in the proportion of 28 and 72 for 100 atoms). It differs from oxygen in that one of the isotopes is radioactive and decays—as we saw with strontium-87, which is the basis of one geological chronometer.

These two simple examples permit us to address some fundamental questions. In the case of oxygen we see that one isotope (^{16}O) is much more abundant than the others. Nature has recipes for making elements, and for oxygen it is clear that the recipe is much more efficient for one type of nucleus than for the other. On this basis we can see that if the proportion of isotopes of the same element varies, these variations must reflect differences in the natural fabrication process. Are the isotopes all made at once? If so, their proportions reflect variations in the physical parameters of the fabrication process. Or are they made "independently" and then mixed later in interstellar space? The different proportions would then reflect variations in the mixing conditions. Isotopic variations carry messages about the fabrication process within themselves. For rubidium the initial mixture is certainly modified by the radioactivity of the rubidium-87 isotope. Radioactivity is a process that destroys nuclei, or, more exactly, transforms them into other nuclei. If rubidium-87 still exists, it is because its decay rate is very slow compared to what it was at the time it was formed. Since rubidium's decay period is measured in billions of years, it is clear that the fabrication of elements is a very ancient process. If we knew in what proportion the rubidium-87 isotope was formed, we could calculate the age of rubidium by measuring the abundance of the current residual amount. Is this a fantasy? We will see that it is not.

There is one further issue we must examine before approaching the processes of nucleosynthesis. A nuclear reaction is to the nucleus what a chemical reaction is to atoms. Two nuclei can react with each other. From this collision new nuclei (and therefore sometimes new elements) can be created. Having said that, let me try to be more precise.

In certain circumstances the carbon-12 nucleus (^{12}C) reacts with helium-4 (4He) to produce oxygen-16 (^{16}O):

$$^{12}C + 4He \rightarrow {}^{16}O$$

just as in the notation of chemical reactions we write:

$$C + O_2 \rightarrow CO_2.$$

(Note that no new isotope appears in the product of a chemical reaction, whereas in a nuclear reaction the object is precisely that—the fabrication of a new isotope.) Setting off ordinary chemical reactions necessitates overcoming the mutual repulsion of the external electrons. To ignite the explosive chemical reactions of a gas motor, a fire must be started by a spark. Nuclear reactions (alchemy) take place, as we already know, only in very specific and difficult-to-achieve conditions. The atoms must be raised to very high temperatures on the order of millions of degrees, because the mutual repulsion of the nuclei is much stronger than that of electrons. Setting off a nuclear reaction necessitates first overcoming the repulsion among nuclei, then penetrating this forbidden inner space and triggering a redistribution of the protons and neutrons. This triggering or igniting benefits from the tunneling effect discovered by Gamow. It is easy to understand that the heavier the nuclei, the larger the number of protons and the higher the charge and the mutual repulsion. The energy necessary to trigger the nuclear reactions of heavy nuclei is greater than that necessary for light nuclei. Hydrogen nuclei react more easily than oxygen or nitrogen nuclei and much more easily than lead nuclei. But as in an explosive chemical reaction, once the nuclear reaction has started it gives off heat. This heat or energy is much greater than that needed to start the reaction, so a nuclear reaction is a real energy amplifier. It uses matter and turns it into energy several thousand times more efficiently than any chemical reaction. This is why nuclear power plants consume much less uranium than conventional plants do coal.

The relationship between nuclear reactions and energy is fundamental in all that we are addressing here. It is fundamental because it links astrophysics and nuclear energy, but more especially because the equivalence of matter and energy expressed by relativity takes all of its meaning on this level. Particles and bonding energy are equivalent notions, and conservation of mass and conservation of energy are henceforth equivalent laws.

The Steps in Nucleosynthesis

The nuclear game is a sort of building game. Starting with hydrogen—an atom consisting of one proton with one electron circling around it—we add a neutron (a proton that has captured an electron) and build the different nuclei of the universe onto this base through successive additions, which take place during nuclear reactions. This game would be simple, even elementary, if the laws of nuclear physics did not get in the way.

Hydrogen Fusion and the Creation of Helium

The first problem to solve is the most difficult, not to achieve but to understand. How do we get from hydrogen to the second element, helium, whose most abundant isotope has a nucleus consisting of two protons and two neutrons? The reaction, illustrated in Figure 36, takes place in steps (the first two steps consist of adding a hydrogen nucleus), creating first an isotope of hydrogen, deuterium (by expelling a positron), and then an isotope of helium, 3He (two protons, one neutron), along the way. This step is followed by a fusion of two nuclei of 3He, which expel two hydrogen (protons) and produce a nucleus that is particularly important for the following operation, helium-4 (4He). Its symmetry of construction (two protons plus two neutrons) gives it a remarkable stability, which makes it an essential brick in the process of building nuclei and therefore elements.

The Multiplication of Helium-4 (Element a)

The next step is that of the fusion of helium nuclei multiplied by 3, 4, 5, 6, 7, 8, 9, 10. This produces nuclei of masses 12, 16, 20, 24, 28, 32, 36, and 40, all consisting of an equal number of neutrons and protons, all very stable and very abundant compared to their neigh-

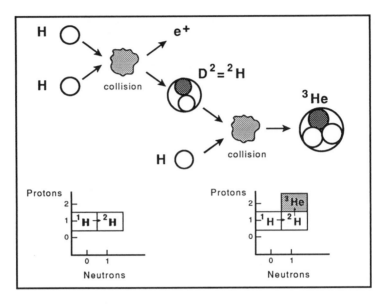

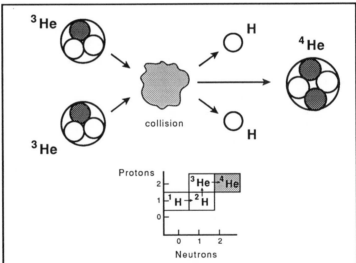

Figure 36 The early stages of nucleosynthesis, starting with hydrogen (top) and ending with helium 4 (bottom), illustrating the relevant sections of the protons-neutrons diagram.

bors. In fact, they correspond to the most abundant isotopes of the following elements: carbon, ^{12}C; oxygen, ^{16}O; neon, ^{20}Ne; magnesium, ^{24}Mg; silicon, ^{28}Si; sulfur, ^{32}S; chlorine, ^{36}Cl; and calcium, ^{40}Ca; all of which are fundamental in our world.

Readers will have noted, however, that I have not mentioned an element that would correspond to the addition of two helium nuclei of mass 8, which would be berylium-8. This element is unstable. In fact, the instability of isotopes with a mass of 8 and the way that carbon is built from helium constituted one of the great puzzles of nucleosynthesis solved by Edward Salpeter. Berylium-8 is unstable, but it forms relatively easily. It therefore provides a bridge, a passageway or transitory state by which to pass from one 4He to three 4He (that is, ^{12}C).

In nature, this process of adding successive building blocks of 4He one by one can be skipped over by constructing "prefabricated" blocks. Addition is outstripped by multiplication! Thus, to obtain magnesium-24, there is a slow method, which is to add successively six nuclei of 4He, and a fast method, which is to add two carbon atoms:

$$^{12}C + {}^{12}C \rightarrow {}^{24}Mg.$$

This latter method is called *carbon fusion.*

In the same way, we can fuse two superbricks of oxygen to obtain sulfur:

$$^{16}O + {}^{16}O \rightarrow {}^{32}S.$$

These two short-circuit operations (see Figure 37), especially the first two, play a fundamental role in the synthesis of elements because they are faster, and therefore more efficient, than the laborious addition of successive helium-4s.

An examination of the table of chemical elements or that of the isotopes shows us that with the process derived from the 4He blocks (or from the superblocks of ^{12}C or ^{16}O) alone, we cannot obtain all the elements. We have not produced nitrogen-14 or fluorine-18 or sodium-22 . . . or the minor isotopes of oxygen-17 and oxygen-18 we have talked about, or carbon 13 . . . For that we have to use hydrogen reactions. Thus ^{12}C $+ {}^{2}$H $\rightarrow {}^{14}$N (nitrogen) and ^{20}Ne $+ {}^{2}$H $\rightarrow {}^{22}$Na (sodium), but those reactions are less efficient than 4He

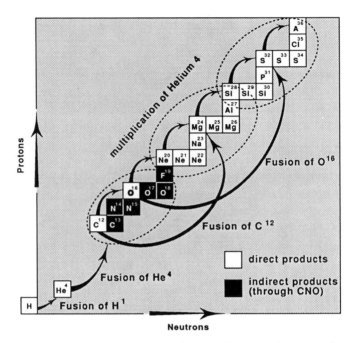

Figure 37 An illustration of helium, carbon, and oxygen fusion on the protons-neutrons diagram.

additions, and the products are thus less abundant than their neighbors.

From Nuclear Chemistry to Astronomy

To explain the abundance of the chemical elements in the universe by using the little game of nuclear building blocks would be the purest speculation if it were not solidly supported by other findings. As we have already noted, nuclear reactions take place only when the nuclei themselves are sufficiently "hot," that is, sufficiently agitated, to collide with each other at very high speed. A temperature of a million degrees is necessary to set off hydrogen fusion. It takes 100 million degrees to fuse helium nuclei and 600 million to fuse carbon and oxygen.

These conditions can be obtained in the laboratory in particle accelerators, in which the characteristics of all these reactions, their greater or lesser ease of being produced, can be measured as probabilities. At the same time, the temperatures necessary for these reactions and the heat given off by them can also be measured. This information prepares us to calculate the characteristics of the fusion reaction we have hypothesized. And yet the problem remains theoretical—a theory refined and supported by experience but still a theory. In nature where can temperatures high enough to set off nuclear reactions be found?

As we know from the work of Atkinson and Houtermans, natural nuclear reactions occur in the stars. The stars are gigantic nuclear reactors whose luminosity is maintained because of the formidable quantity of heat emitted by the fusion of atomic nuclei. The light the stars produce makes it possible to determine the surface temperature (color), the amount of energy radiated (intensity), and, thanks to spectroscopy, the chemical composition of each of them. It is therefore possible to learn which type of nuclear reaction takes place in which kind of star. Our sun is continuously fabricating helium because its temperature of a few million degrees permits only hydrogen fusion. Helium fusion takes place only in the larger stars, called *red giants,* the fusion of oxygen in *supergiants.* The stars are thus nuclear reactors whose high temperatures set off nuclear reactions that in turn give off heat as they occur. Since the stars obey the principle of conservation of energy, they are evolving (see Figure 38). They have a history: they are born and they die. The life and death of a star: a beautiful scenario in perspective.

Let us imagine a hydrogen cloud. If it is massive enough, it contracts through the force of gravitation. During the contraction the particles collide with each other and heat up. When the temperature reaches a million degrees, the fusion of hydrogen begins, emitting considerable heat. The ball of gas becomes a star and shines brightly. But the star loses energy by emitting light. An equilibrium is established between the energy created in its interior through nuclear reactions and its brightness, which is the visible sign that it is dissipating that energy. A moment will arrive when the reactor can no longer function for lack of fuel (our sun will reach this stage in five billion years). At that point the star contains many helium nuclei. Since each helium nucleus consists of four hydrogen nuclei, the density of the gas has become lower. The star therefore again contracts under the effect of gravitation, its temperature again increases, and new nuclear reactions, more difficult to initiate, are set off. So it is

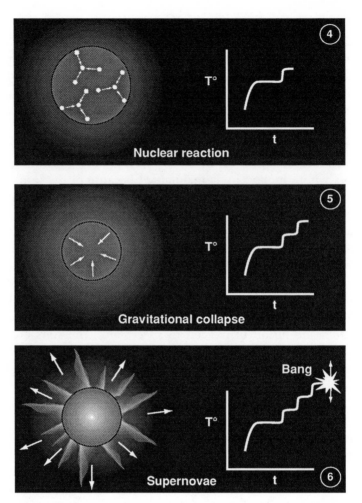

Figure 38 The final stages of star evolution, showing the alternating gravitational collapse and nucleosynthetic episodes. On the right are temperature/time graphs.

with the fusion of helium at a temperature of 100 million degrees. But at that point the star is no longer a sun; it has arrived at the red giant stage.

Earlier in this century, Ejnar Hertzsprung and H. N. Russell, both astronomers, classified the stars according to their luminosity and their temperature on a graph now called the Hertzsprung-Russell (or

H-R) diagram (see Figure 39). All the stars that fuse hydrogen are found on the same diagonal band in the diagram, which defines what astronomers call the *main sequence*. Besides our sun, many of the well-known stars, such as Alpha Centauri, Sirius, Castor, and Vega, are also found there. The red giants are outside the main sequence,

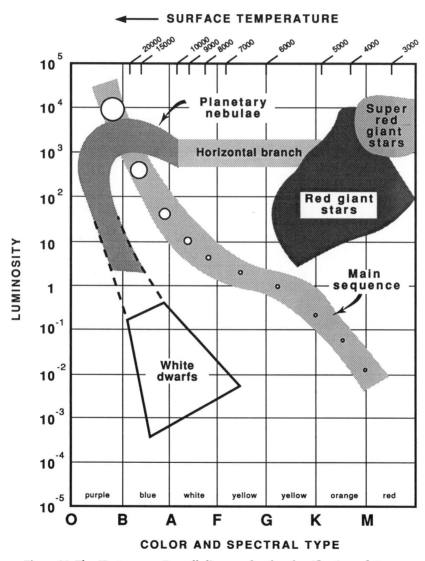

Figure 39 The Hertzsprung-Russell diagram for the classification of stars.

in the upper right of the diagram. Quantitative models built by nuclear astrophysicists have made it possible to explain the distribution of stars in the H-R diagram and thus to determine which steps of nucleosynthesis take place in which types of star. We can follow the trajectory of a star in an H-R diagram during the course of its evolution, for example, its passage from the main sequence to the red giant zone or from the prestellar cloud to the main sequence. Our game of nucleosynthesis has now been justified since it not only explains the abundance of elements and isotopes but does so quantitatively, and integrates in a comprehensive model both the experiments of nuclear physics and astronomical observations—an impressive synthesis indeed.

But we have not yet explained everything, in particular how the elements heavier than sulfur, those whose atomic numbers are greater than 32, are fabricated, or how the stars die.

The Iron Family

Above mass 40 the game of adding helium blocks reaches its natural limits. In fact, at that point nuclear stability moves away from the principle of an equal number of protons and neutrons. To assure the stability of the nucleus it is henceforth necessary to have more neutrons, more "cement," because the mutual repulsion of the protons becomes greater.

In the same way the mutual repulsion of heavy nuclei becomes very great and to overcome it, to initiate nuclear fusion, extremely high temperatures, around 5 billion degrees, must be reached. At that temperature the collisions are very violent and as many isotopes are destroyed as are created. Also at that temperature light itself reacts with nuclei and destroys them (photodisintegration). Instead of constructive nuclear reactions we are watching a nuclear snowball fight in which the chance of breaking nuclei into simpler nuclei is high; from time to time, in a happy accident, a slightly heavier nucleus is formed.

In this statistical equilibrium, the survivors are helium-4, the always stable superblock, as well as the nuclei whose nuclear stability is the greatest, that is, those whose interior construction allows them to resist destruction. The winner is iron, with its principal isotope, ^{56}Fe. It has the highest binding energy per nucleus. Iron's neighboring elements are also very stable, so various isotopes of titanium, vanadium, chlorine, manganese, cobalt, nickel, and zinc are synthesized as well.

After Iron: The Heavy Elements

If no element heavier than zinc or copper existed, nuclear physicists could easily justify that fact. To make heavier elements through nuclear reaction the firing temperature must be raised ever higher to overcome the repulsion among very heavy and heavily charged nuclei. If the temperature is increased to five billion degrees, calculations show that destruction wins the snowball fight over construction. Trying to construct nuclei heavier than zinc or copper by mutual bombardment, we destroy them and fall back on . . . helium-4! In nature, however, elements heavier than iron do exist: gold, silver, platinum . . . all the way up to uranium, which is present but unstable. How, then, are they made? We must leave the nuclear fusion method and substitute a more subtle, less violent game.

Neutron Addition

The neutron is electrically neutral. It is not subject to the electrical repulsion of the nucleus. Launched with sufficient energy (but not too much) it can mix into the nucleus, bind to it, and increase its mass. At first sight, an increase in the number of neutrons seems only to increase the number of isotopes of an element such as copper or zinc. How can new elements be made?

At this point radioactivity intervenes. The nucleus can accept only a few supplementary neutrons; after that it becomes unstable and disintegrates. By emitting an electron and therefore transforming a neutron into a proton, it creates a *new element* (remember the path of radioactivity in Figure 34).

Thus, from the addition of a neutron to initiate radioactivity we progress through the periodic table in a zigzag, respecting, of course, the rules of nuclear stability—the laws that establish the acceptable proportions of protons to neutrons for a given number of protons. This mechanism of nuclear creation, which is based on the combination of building by adding neutrons and beta radioactivity, in contrast to the table of natural abundances of isotopes, complicates the process a little and causes the flux of neutrons to enter the picture. When the flux of neutrons is slow and the neutrons are added one by one, natural radioactivity can be initiated (see Figure 40). The first isotopes rich in neutrons and radioactivity are therefore barriers to the production of the other heavier isotopes that could theoretically be formed.

Let us suppose on the other hand that the rain of neutrons is rapid (see Figure 41). Now the radioactivity of the first radioactive isotope

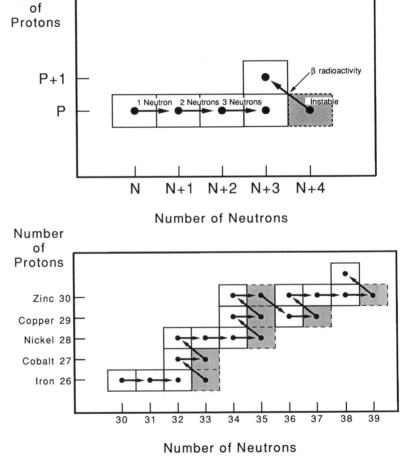

Figure 40 S process illustrated in protons-neutrons diagram. The upper part illustrates the principle; the lower part gives an example.

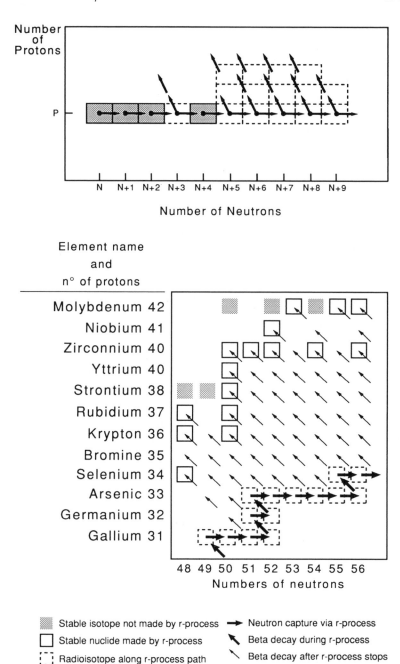

Figure 41 R process illustrated in protons-neutrons diagram. The upper part illustrates the principle; the lower part illustrates an example.

is not enough to stop the horde of neutrons. Certain radioactive nuclei have new neutrons grafted onto them even before being transformed into a new isotope, which is heavier but still the same element. In general, these heavy isotopes are radioactive; they therefore decay into a new element, in this way creating a series of radioactive cascades. Some isotopes that are rich in neutrons are stable, however, because of the rules of nuclear stability. The processes of rapid neutrons are therefore able to fabricate them. This explains how the isotopes rich in neutrons, which exist for almost all the heavy elements, are made.

The process that adds neutrons slowly is called process *s*, the one that adds them rapidly, process *r*. These two processes explain the synthesis of heavy elements up to uranium, which is radioactive and produced only by *r* processes.

The table of nucleosynthesis just outlined should be complete. The way various isotopes of the light elements, such as ^{17}O and ^{18}O and ^{21}Ne and ^{22}Ne, are fabricated remains very obscure to me, even if astrophysicists can explain everything. Proton reactions could be used, and it is always possible to construct explanations case by case, but is this the right way? All these isotopes are rich in neutrons compared to the abundant isotope made of helium blocks or of helium plus deuterium. Is this merely chance? Do we need to enlist neutron synthesis? The theory is not yet established. Whatever the case, these reactions are undoubtedly difficult and unusual, so that the isotopes formed by these secondary processes are rarer.

The Abundance Curve of the Elements

For now, let us go back to the abundance curve of the chemical elements (or the principal isotope of the element, which amounts to the same thing). The abundance curve can be divided into four parts:

- The light elements, hydrogen and helium.
- The elements synthesized by the multiplication of helium (element a) and associated with those for which hydrogen is added.
- The iron peak resulting from statistical equilibrium.
- The heavy elements formed by neutron addition.

The abundance of elements decreases with their mass, except where the iron peak is concerned (see Figure 42).

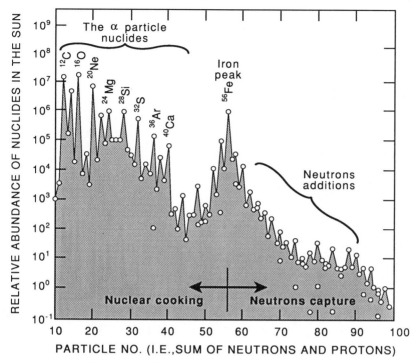

Figure 42 The abundance of nuclides in the solar system showing their different modes of nucleosynthesis.

The first elements—hydrogen, helium, carbon, nitrogen, and oxygen—are the principal constituents of cosmic matter, of the galaxies, stars, and giant planets. These elements also constitute living matter and organic molecules. There is no reason to be surprised by the radioastronomers' discovery that organic interstellar molecules exist in abundance. If we relate the abundance of a cosmic structure directly to the abundance of its constituent elements, must we then conclude that life is an abundant phenomenon in the universe? Is this reasoning justified for an assemblage as remarkable as life?

As Figure 43 suggests, the elements that constitute the inner planets include oxygen, magnesium, silicon, sulfur, calcium, all the a elements, iron, and nickel, with somewhat less sodium, aluminum, phosphorus, and potassium, which are alpha elements. The heavy elements are all present in cosmic objects as trace elements. This table is complete and accurate enough, but there is an important omission: the three elements lithium, beryllium, and boron follow helium

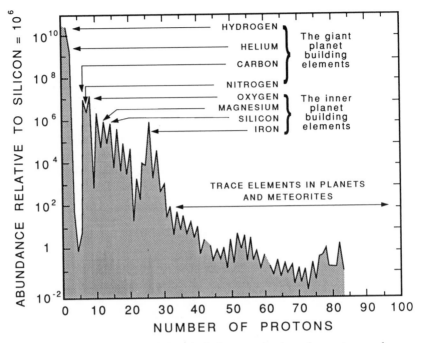

Figure 43 The abundance of chemical elements in the solar system and their preferential distribution in the planets.

immediately, but, like the heavy elements, their abundance is very low. Why?

The answer to this question is, in essence, nuclear: their nuclei are among the least stable, and they are therefore difficult to make and easily destroyed. Nevertheless, their abundance is not zero. Their formation is a subject of debate among astrophysicists. The general opinion is that they are formed by secondary reactions due to the cosmic galactic rays in interstellar space, but the data to substantiate this hypothesis are certainly far from complete.

The Stellar Scenario

If, as we have seen, the elements are synthesized in the cosmos, we then ask, where are the heavy elements formed?

The red giants are the site of important reactions. All the alpha reactions take place there, but not randomly. This is the case for Sirius in the constellation Canis Major, Vega in Lyra, and Rigel, Betelgeuse, and Bellatrix in Orion's belt. Like most celestial bodies, the stars have the layered structure of an onion: the temperature at the center is much higher than that at the surface, so successive layers are superimposed. In each layer, a series of nuclear reactions takes place. The statistical equilibrium of the iron peak occurs at the center of the red supergiants, the fusion of carbon and oxygen in the mantle.

Why are the centers of the stars hotter than their exteriors? For two reasons: first, the exteriors lose energy through radiation (the sun cools off as it warms), whereas the centers do not; and second, the forces of gravitation cause the atoms in the interior to collide and heat up, and this temperature elevation sets off nuclear reactions that give off energy, which essentially remains trapped.

No place on the normal H-R diagram, however, seems suitable for synthesizing very heavy elements through the addition of neutrons. Neutrons are interesting particles for nuclear reactions, but they are difficult to produce as free agents. Astrophysicists believe that the sites of such production are short-lived and rare: they are the *supernovae*.

If a star is large enough after the red giant stage, it can eventually set off a new phenomenon, complex and spectacular and extraordinary: the explosion of a supernova. Within a few minutes the star reaches a temperature of dozens of billions of degrees and explodes. This explosion projects debris of all sorts—the chemical elements it fabricated and contained—into the universe in every direction. During this time its center becomes dwarfed and gives birth to a neutron star or *pulsar*. This phenomenon, observed by the Danish astronomer Tycho Brahe in 1572, was studied recently thanks to the explosion that took place 170,000 years ago in the Andromeda galaxy in the larger of the Magellanic Clouds, whose light finally reached us in 1987.

Astrophysicists believe that heavy element synthesis takes place when red giants form supernovae. For them the processes of slow neutron synthesis *(s)* take place in the envelopes of the red giants following complex pulsations, while the rapid neutron synthesis *(r)* occurs exclusively in the supernovae.

In fact, not much is known about these syntheses. One reason is that we lack direct observations. The events of 1987 are therefore very important. All the observations are not yet completely analyzed, but the detection of neutrinos seems to be clear and reinforces the general scenario. Observations continue, especially in X rays and gamma rays. Also, since the scenarios vary with the mass of the star, the theory too can vary.

The other possible end of a star is less glorious. Lacking fuel, it sinks into itself without an increase in temperature and it becomes a little *white dwarf* before imploding in a relatively discreet way. Born in a cloud of interstellar gas, the star dies by exploding, creating a cloud of gas. It is almost a cyclical process. But the gas cloud created by the dying star is enriched with the heavy elements the star was able to fabricate. This second generation cloud mixes in space with other clouds that have other histories and other chemical compositions. These crossbred clouds become concentrated at some point, transform themselves into a star, and the whole process repeats itself.

The universe evolves from mixtures of explosion-produced clouds continually enriched with new heavy elements. The stars are the high point in the history of this universe of interstellar clouds, and the supernovae their moment of reproduction. They inject their seed into the universe, ready to engender other stars.

Our sun is a star of the second or third generation. Spectroscopic analysis of its light tells us that it contains all the chemical elements with which we are acquainted. Its abundance of heavy elements—iron, for example—is as great as that of the interstellar clouds that surround us. We know that the sun itself can make only helium. Its other elements are inherited from its previous history. They are pre-solar elements, antique elements born in other stars—in red giants and in supernovae that exploded at a time when neither sun nor Earth existed. Those ancestors were numerous, and their elements have been mixed together many times. Of course, this scenario applies to the Earth as well, and to all the elements in the solar system. The carbon in our bodies, the oxygen, silicon, and magnesium in rocks, the iron in the terrestrial core, and all the other elements are primordial.

All the elements that constitute our bodies and the objects that surround us were fabricated in the stars billions of years ago, before the sun and our planets were born. In five or six billion years these elements and, therefore, we ourselves, will return to the cosmos and eventually, no doubt, to a new star. Thus the universe evolves in successive cycles, becoming progressively enriched with heavy elements. But how did all this begin?

The Big Bang and the Initial Synthesis

The theory of the synthesis of elements in the stars (called *astration* theory) explains most of the observed abundances very well. But it does not explain everything. It does not explain the abundance of the two near neighbors of hydrogen, helium-4 and deuterium. In any region of the universe, helium-4 is always ten times rarer than hydrogen. If helium is formed, as is thought, in the stars of the main sequence, its abundance should vary with the direction of its source, with the possible frequency variations of one or another scenario, as the abundances of the other elements vary. But it is constant. Does this mean its constance is primordial? As for deuterium, which is so vulnerable to the absorption of neutrons and to thermonuclear fusion with protons, its abundance is too great. It should not exist, or barely, since its combustion in the stars occurs more quickly than its production. Since its present abundance cannot be explained by astration theory, could it be primordial as well?

The two fundamental questions posed by the existence of this contradiction between theory and observation led astrophysicists interested in the synthesis of elements to become interested in the Big Bang. The scenario for the first "instants of the universe" came out of astronomical observations, notably the recession of the galaxies and the existence of the fossil three-degree radiation. The theory has been carefully developed and acknowledges the most up-to-date understanding of particle physics.

The scenario used by the astrophysicists has made it possible to understand appropriately the abundance of helium-4, deuterium, and even the formation of the fundamental element, hydrogen. From a primordial universe of elementary particles hydrogen atoms were born during a gigantic expansion that projected matter toward the four corners of the universe. Tracing the recession of the galaxies backwards, back to their beginning, astrophysicists place the initial instant between 10 and 20 billion years ago. The uncertainty about whether this is truly an "initial" instant comes from the fact that it is not known whether the universe is infinite or closed and whether, after its current period of expansion, it will recontract. In the present state of our knowledge (or rather, our ignorance) it is thought that the Big Bang marks time zero of "our time." It is the basic reference point, just as the birth of Jesus Christ is for the Christian calendar.

The Isotopic Palimpsest

An essential moment in our journey occurs here.

We have seen that the data of nuclear physics, the observation of the stars, and theoretical calculations produced a general scenario for nucleosynthesis, that is, for the fabrication of the chemical elements. Variations on the standard scenario are possible but not proven. Did all stars pursue the same evolutionary cycle? Do their chemical and isotopic compositions differ or not? Not all interstellar space is homogeneous, and mixtures of diverse origins vary. The relative abundances of the various isotopes of the same element and of the various elements in the universe are not fixed a priori. Everything indicates that important variations in the scenario, and therefore in isotopic and chemical abundances, can exist. Yet some would say that everything is uniform. Where is the truth?

The measurement of the isotopic compositions of the natural elements, like the measurement of the abundance of these elements, is a constraint and a test for all the scenarios of nucleosynthesis and the synthesis of elements from which our solar system, and we ourselves, came. A variation in these compositions indicates a fluctuation in the scenario that has to be explained and whose mechanisms have to be determined. But an earlier question returns. Are the measurements of the isotopic composition and chemical abundances of complex planetary objects as reliable as we can obtain? Is the evidence received in this way accurate? Hasn't the composition been altered by time? The answer to these questions is important. In other words, if we observe chemical or isotopic variations among natural objects, can those variations be attributed to nucleosynthesis? Or are they primordial?

For chemical abundances the answer is a little complicated. Certainly many of the large differences in concentration that we observe in planetary objects, on Earth, or in meteorites are inherited from nucleosynthesis.

If uranium is measured in parts per million in the terrestrial crust while silicon is measured in percentages, this is no doubt a result of nucleosynthesis. But this heritage becomes obscure as soon as we try to systematize it or to explain the differences in a detailed way. A gold mine cannot be explained by nucleosynthesis.

Unlike chemical composition, isotopic composition goes through planetary events almost unaltered. The average isotopic composition of the iron in an interstellar cloud is identical to the composition of the metallic iron in the meteorites that come from it, just as the isotopic composition of the iron in an olivine crystal is identical to that of the magma from which it crystallized.

To a first approximation, the isotopic composition of the majority of elements is conserved throughout the vicissitudes of planetary life. Chemical processes do not divide the isotopes among them, since they are atoms of the same chemical element. We can therefore expect the isotopic analysis of a chemical element taken from a planetary object (gas, liquid, or solid) to reveal the way in which it was formed and its nucleosynthetic history. Encoded in its isotopic composition the message of nucleosynthesis is preserved through time despite the fluctuating conditions of cosmic life. Its isotopic composition can be modified only by a nuclear reaction, including radioactivity, or by mixing it with an element that has different isotopic composition. Underneath the complex, varied, and variable chemical message an isotopic message that is practically inalterable lies coded. More robust and simple, it records in the very heart of the atom the signature of its origins. Nothing erases anything.

The Age of the Elements

I have already alluded to the idea that the isotopic composition of a radioactive element could be used to calculate when it was formed. If, by using the concepts of nuclear physics, we determine the original isotopic composition—that which was produced during nucleosynthesis—we can calculate the time that was necessary to obtain the present isotopic composition by using the principle of the hourglass.

This principle has been applied to uranium. As we know, this element has two natural isotopes, uranium-235 and uranium-238, both unstable, both radioactive. Their periods of radioactive decay are different, uranium-235 decaying much faster than uranium-238. The problem is that the isotopic ratio of ^{238}U to ^{235}U today is 137/8.

According to astrophysicists the isotopic ratio produced during nucleosynthesis was around one (0.6 to 0.7). What then is uranium's age?

Suppose that all the uranium in the universe was produced at once, in a single event (a single supernova). The calculation for the age of uranium would be 7 billion years, hardly more than the age of the solar system (4.5 billion years). This scenario of a sudden unique creation is not very likely. We know that there have been thousands of explosions of supernovae (in our galaxy one about every hundred years).

It is more reasonable to construct another scenario and to hypothesize that uranium was formed continually (but not uniformly) over the course of cosmic time. If this was the case we can calculate an average age for uranium of 13 to 15 billion years.

The same calculation used for uranium can also be used for rhenium. This element has an isotope, rhenium-187, that decays to osmium-187. The isotopic composition $^{187}Os/^{186}Os$ produced in nucleosynthesis by the *s* process is well known. Using the ratio $^{187}Os/^{186}Os$ at the time of meteorite formation it is possible to estimate how much osmium-187 is due to the radioactivity of rhenium-187 in the presolar stage. The measurement of the Re/Os cosmic ratio makes it possible to estimate an age for the formation of rhenium of 13 to 20 billion years.

The calculations show agreement. Can others be made? One would hope so, but we are limited by our knowledge of the nucleosynthetic process. In other words, we don't know the exact initial isotopic composition of the other elements with long-lived radioactive isotopes. We must therefore content ourselves with these two calculations, to which two other less reliable ones can be added, those for the $^{232}Th/^{238}U$ and the $^{87}Rb/^{87}Sr$ ratios, which also furnish figures of around 15 billion years.

What exactly does this age mean? It is an average formation age for the heavy elements, a sort of weighted average of the temporal distribution of supernovae over the course of cosmic time. This age takes on additional significance when compared to the age determined for the Big Bang using the recession speed of the galaxies or of the globular clusters, which is also close to 15 billion years. Their agreement indicates that stellar activity, the formation of stars and the explosion of stars in the form of supernovae, has actually been rather uniform and continuous over the whole course of cosmic time.

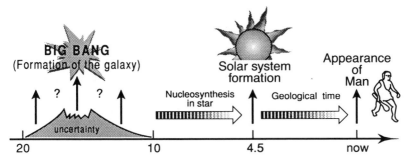

Figure 44 Cosmochronology (billions of years).

The decrease in activity was not abrupt; otherwise the scenario for the secondary formation of uranium would have furnished an age of 15 billion years rather than 7. The synthesis of the heavy elements in the *r* process—the successive activity of the supernovae—has been maintained over time. It has certainly decreased, but not too dramatically, and constitutes a quantitative measure of the evolution of our universe that is of the highest importance.

Such a scenario seems dubious to astronomers, who have a tendency to think, in their great cosmic vision, that the activity of the universe decreased rapidly after the Big Bang. A new estimate of the age of the Big Bang at about 10 billion years, as William Fowler proposes, would resolve the contradiction. But are these new determinations justified? This question brings us to the very frontiers of science, and it would be premature to attempt conclusions at this time.

From another point of view, this scenario gives us a cosmic calendar, which provides our terrestrial chronology with a broader context (see Figure 44). The ancients confused the births of the universe, the Earth, and Man, placing them in a single cosmic instant. The radioactive clock lets us see that these events were distinct, that it took an evolution of 5 to 10 billion years to go from the Big Bang to the solar system and another 5 billion years to arrive at Man. The cosmic calendar puts the major events of our origins in perspective.

The History of the Oxygen Isotopes

The history of science is full of the unexpected, and the way in which ideas are born and develop does not always follow the "Cartesian" model. At the end of the 1970s the future of planetary cosmology looked very bright. Thanks to the efforts of researchers such as William Fowler at Caltech, Fred Hoyle at Cambridge, Al Cameron at Harvard, and a few others, the scenario of nucleosynthesis, as we have examined it schematically, was widely accepted. It explained the chemical and isotopic abundances of the various objects in the universe, the astronomers' observations, and the very natures of the different stars; it also integrated the most up-to-date understanding of nuclear and atomic physics. The theory of the condensation of a hot nebula, as developed by Al Cameron (1970) and Ed Anders (1971), explained the existence of the various chemical compounds present in the solar system. Supported by the theory of homogeneous accretion, the theory of condensation in a hot and heliocentric nebula explained the chemical diversity existing in the various planets, the near as well as the distant ones. John Lewis (1973), the young apostle of this synthesis, then at MIT, could conclude that our understanding of the origin of the solar system had made a decisive step and that only a few secondary details still needed to be explained. It seemed that the evolution of the universe from the Big Bang to the formation of the Earth was completely understood.

But this scenario was to receive a shock from new data on the isotopic composition of elements in meteorites. Robert Clayton of the University of Chicago specialized in the study of the abundance of the isotopes of oxygen. Oxygen is the most abundant element on Earth. It is abundant in the air, in water, and also in silicates in which, with silicon, it forms the framework for silicate compounds. As we will see, the isotopic composition of oxygen, especially the abundance of a minor isotope, ^{18}O, compared to that of a major isotope ^{16}O, (the $^{18}O/^{16}O$ ratio) can be used to decipher the intimate mechanisms of natural chemical reactions. For example, the $^{18}O/^{16}O$ ratio in the shells of extinct animals can be used to determine the temperatures of the water in which they lived.

If the $^{18}O/^{16}O$ ratio is an extremely powerful geochemical indicator, it must be admitted that it is also very difficult to measure. The variations observed in the ratio are so small that they are expressed as one part in ten thousand. To make these variations more meaningful, they are measured as deviations from a known standard, that of seawater. Thus, when it is said that a given isotopic $^{18}O/^{16}O$ compound is $+12$, it means that the ^{18}O isotope is enriched by 12 ten-thousandths compared to seawater; a negative number means that it is depleted.

Measuring the isotopic composition of oxygen in rocks, minerals, water, and so on using the mass spectrometer is difficult and demands constant attention, with frequent checking and calibration. During one of these calibrations Clayton noticed an apparently unimportant fact: the oxygen-17 isotope, a forgotten isotope to which no one usually paid any attention, seemed abnormally abundant in some samples. Generally, the deviations in the $^{17}O/^{16}O$ ratio are half those of the $^{18}O/^{16}O$ ratio, which is explained by the fact that the difference in mass between ^{17}O and ^{16}O is only half that of the difference in mass between ^{18}O and ^{16}O. In these "strange" samples Clayton found that the deviations were identical for the $^{17}O/^{16}O$ and the $^{18}O/^{16}O$ ratios. Either ^{17}O seemed to be overabundant or ^{18}O to be underrepresented (which amounts to the same thing). Clayton made more measurements and ascertained that a whole series of rock samples showed this anomaly. All were meteorites, and those with the largest anomalies were carbonaceous chondrites. Using a simple graph Clayton showed that all the observed anomalies could be explained by mixing oxygen of terrestrial isotopic composition (T) with very abnormal oxygen (A) consisting exclusively of ^{16}O (see Figure 45). The differences are due to the fact that the proportions of the mixture vary among meteorites. Some are very rich in oxygen A, while others are less so (Clayton et al., 1976).

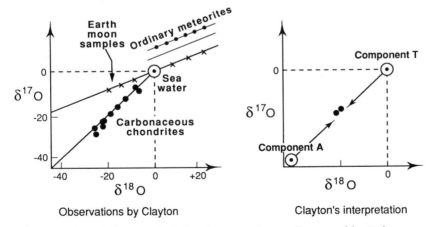

| Observations by Clayton | Clayton's interpretation |

Figure 45 Oxygen isotope variation in meteorites as discovered by Robert Clayton. The left part illustrates the data; the right part illustrates Clayton's interpretation of the data by mixing different components.

Clayton's discovery burst like a bombshell on the scientific world. For the first time isotopic variations that could not be attributed to radioactivity or simple physicochemical phenomena were seen. The isotopic heterogeneities of oxygen are primordial. They are the product of several fabrications of oxygen nuclei. Some stars synthesized ^{17}O and ^{18}O impurities along with the ^{16}O, while others were able to make almost pure ^{16}O through four 4He additions.

As I have said, the idea of variability in the scenarios of nucleosynthesis is not surprising at first glance, and astrophysicists had long admitted such a possibility. More surprising was the existence of such variations among objects belonging to the same stellar system, the solar system. These discoveries did not change the scenario of nucleosynthesis fundamentally; they established its necessary variation around a standard scenario.

Clayton continued his research. He rapidly established that each class of meteorite had a characteristic oxygen isotope signature (Clayton et al., 1976). They were less varied than the signatures of the carbonaceous chondrites (the anomalies were smaller) but differed among the different meteorite groups H, L, and E. Terrestrial samples of this isotopic typology were different from the various chondritic meteorites but had a signature identical to that of lunar rocks, a very strong argument in favor of a common origin for both. Gradually the idea took shape that in the presolar nebula there was a spatial distribution, or cosmography, of the isotopic composition of oxygen. Moreover, the meteorites studied appeared to be homogeneous within each group but heterogeneous as a whole. This posed several questions: Could the asteroid belt be a region in the universe where objects of various planetary origins, from comets to pieces of planets torn off by impacts, have been trapped? Could Jupiter, a great center of gravitational attraction, have been the focus of the formation of a gravitational sink?

Clayton and his group continued working. Dissecting a type 3 (not very metamorphosed) meteorite, they showed that even a single meteorite sample was isotopically heterogeneous. The various pieces of the same meteorite had such different compositions that they must have come from very different places in the nebula. So the problem had to be looked at not on the scale of planetary or meteorite-family heterogeneity, but on that of cobblestone heterogeneity. This time the theory of the condensation of the hot protosolar nebula was discredited. The promoters of the hot nebula theory had assumed that the nebula was set in motion by the violent, turbulent motion of gases, which maintained their homogeneity. If the gas initially consisted of various components of stellar origin, the heat and the violent motion of the nebula had homogenized them all. To support this scenario they pointed to a well-established observation—that the isotopic compositions of the elements were uniform in all the objects in the solar system (Earth, meteorites, the moon, and so on).

The heterogeneities found by Clayton destroyed this whole scenario. How could heterogeneities have existed in such a hot, turbulent environment? And the existence of heterogeneities *within* a meteorite obliged them to go even further: How could the solid grains that constitute the meteorites have come from the condensation of a single gas if they have such isotopic differences? Isn't it more reasonable to assume that meteorites are agglomerated cold from grains and gases of multiple origins and that the solar nebula was a cold mixture of gas and dust, not the incandescent gas that Cameron had postulated?

The idea gained weight from the findings of Gero Kurat (1970), a Viennese mineralogist whose meticulous examination of the Allende meteorite did not support Larry Grossman's (1972) conclusions. Kurat believed that oxides rich in aluminum and titanium, those predicted by the theory of condensation, were not the primary condensates; rather, they were the residues, the minerals that resisted strong secondary heating. For him the composition of the Allende meteorite did not suggest a mineralogical assemblage resulting from cooling but, on the contrary, an extremely heterogeneous assemblage that had been subjected to secondary reheating after its formation. Such an interpretation of textures did not square with the popular condensation hypothesis but suggested instead that meteorites were formed from a cold cloud of gas and dust, which was subsequently heated violently (for example, by the activity of the sun in full growth). This scenario is far removed from those supported by Lewis and others.

Extinct Radioactivities

I have already mentioned the case of iodine-129, a radioactive isotope with a half-life so short that none of it is left today. John Reynolds at Berkeley detected its presence only because its radioactive descendant, xenon-129, was superabundant in certain meteorites. Variations in the abundance of xenon-129 made it possible to construct an extremely precise chronometer for meteorites.

Let us try to use this discovery in the context of nucleosynthesis.

If iodine-129 existed when the solar system was formed, it is because it was fabricated slightly before. Astrophysicists calculated the isotopic ratio of ^{129}I to ^{127}I produced in stars. Then, calculating the same ratio 4.5 billion years ago (by using the excess of measured xenon-129), Reynolds was able to show that the element iodine was formed 100 million years before the formation of the solar system. This result, obtained in the 1960s, was quickly confirmed by the discovery of another extinct radioactivity, plutonium-244, by Paul Kuroda and his students working in Lafayette, Arkansas.

This discovery complicates our chronological table of the synthesis of elements a little. It must be admitted that nucleosynthetic activity took place several million years before the birth of the solar system. The heavy elements have an average age of 10 to 15 billion years, but some of them are much younger. Nucleosynthetic activity was still present a few hundred million years before the solar system formed.

For some reason the work on extinct radioactivities lost some of its vigor in the 1970s (perhaps because of the interruption of lunar exploration). But the discovery of the oxygen anomalies had the effect of a detonator in rejuvenating interest and set off the race to find other extinct radioactivities. If the presolar gas/dust was not very well mixed, as the isotopic variations of oxygen suggested, why not look for other extinct radioactivities, but this time with a shorter lifetime than iodine? Why a shorter lifetime? Because if a rapidly extinct radioactive isotope was present in an object in the early solar system this proves that it was formed very near to the time when the solar system formed and just before those planetary objects were formed. Otherwise it would have died without leaving any imprint. Using the same calculations as for iodine, one can perhaps demonstrate nucleosynthetic events even closer to the formation of the solar system.

The winner of this race for extinct radioactivities was incontestably Jerry Wasserburg and his team at Caltech, and his success did much to win him the Crafoord Prize for 1986. His first and no doubt most important discovery was aluminum-26. This radioactive isotope has a very short lifetime (0.73 million years) and decays into magnesium-

Gerald Wasserburg of Caltech, the pioneer of isotopic studies in extraterrestrial materials. He received the Crafoord Prize in 1986.

26 (see Figure 46). It was in studying the isotopic variations of magnesium in the Allende meteorite that Wasserburg showed the past existence of aluminum-26. Following the same method as had Reynolds for the iodine-xenon couple, he demonstrated that the abundance of the magnesium-26 isotope varied and that these variations were linked to the aluminum-magnesium ratio. Continuing this research, he was able to find a second short-period extinct radioactivity, palladium-107, whose decay product is silver-107. More recently, at the University of Paris, Jean-Louis Birck and I discovered another extinct radioactivity, manganese-53, which decays into chromium-53 with a half-life of 3.7 million years.

These discoveries are essential because they show that some elements were synthesized less than a million years before the formation of the solar system. These elements do not form randomly: it takes the explosion of a supernova to form palladium-107, manganese-53, and aluminum-26. A supernova that exploded just before the formation of our solar system? What a coincidence! And if this supernova were in fact responsible for the formation of the solar system? And the shock wave that formidable explosion engendered set off the

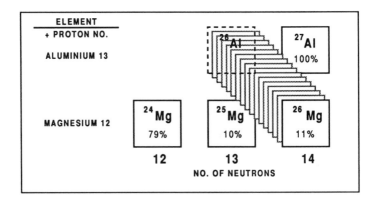

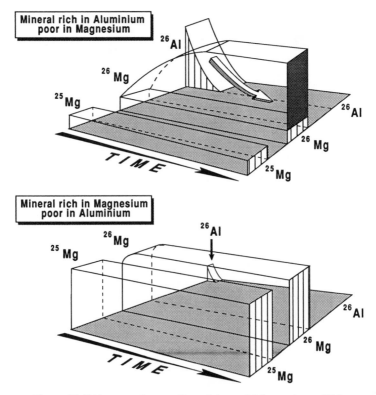

Figure 46 [26]Al, an extinct radioactivity, which produces [26]Mg.

mations. It is as if astrophysicists could link a type of nucleosynthesis and a type of star.

Isotopic Tracers

The isotopic composition of the elements is conserved during complex physicochemical processes. We have seen how useful this astonishing property is in locating the vestiges of nucleosynthesis in the heart of meteorite minerals. The same principal isotopes of certain heavy elements can be used to trace their geologic history, although here the message is not in nucleosynthesis but in the great geologic phenomena translated through the isotopic signature.

The key is long-period radioactivity. We have used it to date rocks and minerals and to determine the age of the Earth, the geologic calendar, and the age of the elements. Now we are going to use its properties to track the great geologic phenomena.

Take, for example, rubidium-87. In the terrestrial mantle, as in other geological reservoirs, it decays into strontium-87. The new strontium-87 is mixed with the "primordial" strontium that has existed in the mantle since the beginning. Because the primordial strontium is abundant enough compared to rubidium, the perturbation introduced into the isotopic composition of strontium (measured, for example, by the strontium-87/strontium-86 ratio) is relatively weak. To be more precise, in 4 billion years the ratio varied from 0.700 to 0.703.

The situation in the continental crust, on the other hand, is completely different. Since rubidium is very abundant compared to strontium in this reservoir, its radioactivity changes the isotopic composition of strontium enormously. In 4 billion years the strontium-87/strontium-86 ratio varied from 0.700 to 0.790 (thirty times greater than that of the mantle).

Suppose that we have a rock whose origin is unknown. Does it come from the crust or the mantle? Measuring the isotopic composition of its strontium gives an immediate indication. If the ratio is low, it comes from the mantle. If it is high, most likely it comes from the continental crust. Moreover, we can also say that its reservoir is probably no more than about 1.2 billion years old (compared with 4 billion years for the high ratio 0.790), and that this age represents condensation of the protosolar cloud? That is the scenario Dave Schramm of the University of Chicago has proposed to explain the formation of the solar system.

But if this scenario is correct, how can we include the existence of the extinct radioactivities, iodine and plutonium? They could not have come from this same supernova, since they are almost 100 million years older. So?

Hubert Reeves of the Centre National de la Recherche Scientifique in Paris, whose scientific activity is as great as his talent as a mediator, has proposed an intriguing scenario. To understand it let us move back a bit.

Paris is famous for its squares, where automobile traffic converges and driving is difficult. There is one square, however, that is less difficult than the others, the Place de la Concorde. In effect it is regulated by two traffic lights. As a result, the flow of traffic moves continuously around the square, constantly renewed, but at the two traffic lights, the concentration of cars is very dense, much more dense than elsewhere. Suppose for a minute that the cars are stars and the square is our galaxy. The stars are indeed turning around the galaxy. The areas around the traffic lights are what astronomers call the arms of our galaxy. As zones in the universe where the concentration of stars is particularly dense, the arms of the galaxy are critical areas where, for example, many explosions of supernovae take place. Our solar system spins upon itself, but it also turns around the galactic center. Every 120 million years it passes into the arms of the galaxy. Why would it not receive a new nucleosynthetic input at that time? Iodine would be the evidence of the passage before last, aluminum, manganese, and palladium of the last passage. If both Schramm and Reeves are right, we were born in the arms of the galaxy. Something to dream about! Such great results from the analysis of little bits of rock!

Isotopes of the Light Elements

A new episode in the saga of the isotopes began in Paris, where François Robert, Liliane Merlivat, and Marc Javoy (Robert et al., 1979) were studying the deuterium/hydrogen (D/H) composition of carbonaceous meteorites. After a difficult year of perfecting the technique—since hydrogen is a component of water, it is easy to contaminate the experiment accidentally—they discovered that some meteorites are considerably enriched in deuterium. These enrichments are much greater than anything known on Earth and cannot be ascribed to terrestrial or extraterrestrial contamination. They are much greater than anything known for oxygen.

Johannes Geiss of the University of Berne and Hubert Reeves (Geiss and Reeves, 1981) rapidly identified the cause of the variations observed in the D/H ratio of meteorites. As we noted, deuterium was principally produced in the Big Bang, and it tends to be destroyed in the process of stellar nucleosynthesis. Its abundance in the solar system is $D/H = 2.10^{-5}$. The value for the interstellar medium is similar. In any case there are considerable deuterium enrichments for interstellar molecules, reaching as high as 10^{-2} for the D/H ratio.

Geiss and Reeves showed that the deuterium enrichments, both in the meteorites and in interstellar molecules, are not the result of special nucleosynthesis but of chemical reactions called ion-molecule reactions that took place at low temperatures in the interstellar medium. In such conditions the isotopes of hydrogen can separate (fractionate), the heavier deuterium reacting more slowly than the lighter hydrogen. Y. Kolodny, John Kerridge, and Joseph Kaplan of the University of California at Los Angeles completed the work of the French group by showing that the deuterium enrichments are not dispersed throughout a carbonaceous meteorite but concentrated in the organic parts, those rich in carbon, that are characteristic of this type of meteorite (Kolodny, Kerridge, and Kaplan, 1980). This second discovery, confirmed by François Robert and Samuel Epstein (1982) at Caltech, has fundamental consequences that transcend the synthesis of elements.

From 1960 to 1970 a few scientists studied the composition of the organic material contained in carbonaceous meteorites. They showed that it contained very complex molecules, including amino acids, molecules that are the basis of organic chemistry. Of course these complex molecules are not very abundant in meteorites, and their origin through contamination by terrestrial products could not be ruled out a priori. In the 1970s the scientific community leaned toward this explanation, refusing to accept the presence of complex organic molecules in meteorites. How could they have formed? Was this another argument for those who believed in extraterrestrial intelligence and extraterrestrials?

Today, measurements of the D/H ratios in complex molecules attest to their extraterrestrial origin. The discovery of interstellar molecules by the radioastronomers showed that there is no longer any reason to be skeptical. We must note, however, that the molecules in the carbonaceous chondrites are infinitely more evolved and complex than those found by the radioastronomers.

Speculations start running wild at this point. If there were extraterrestrial amino acids, why not DNA? Why not life itself? Could life have started in space on grains of interstellar dust? With the help of radiation from cosmic rays? Couldn't these fertile grains have seeded the planets, some of which (like ours) had turned out to be fruitful? This is an audacious scenario for the origin of life, but an appealing one.

Back to the carbon in meteorites. Direct measurement of the isotopic composition of the carbon in carbonaceous meteorites by R. Pillinger's group at Cambridge and Zinner and Epstein in the United States confirmed the work on hydrogen (Swart et al., 1983). The huge enrichments they found resemble those measured by the radioastronomers in interstellar molecules whose origin is the physicochemical fractionation that takes place in the cosmos at low temperatures.

Thus, the study of the isotopic composition of the light elements opens up a new perspective. It is no longer a question of nuclear reactions that occurred at millions or hundreds of millions of degrees inside giant stars or during their explosions. It is a question of reactions that take place on a few grains of dust, on a few molecules of gas isolated in the middle of interstellar space at a temperature close to absolute zero. Isotopic measurements of extraterrestrial dust grains led to distinctly interesting places. The scenario we can derive for the formation of the solar system, however, is akin to the one that Clayton allowed us to glimpse. The materials in the solar system have been formed not from a hot nebula but from a cloud of cold gas and dust of interstellar origin. Carbonaceous chondrites are the best examples of these materials.

But if the variations in hydrogen and carbon are engendered by chemical reactions, can the isotopic variations in the Allende meteorite also be explained in this way? As far as the heavy elements like calcium, titanium, and chromium are concerned, definitely not: their heavy masses make chemical fractionation on the scale measured practically impossible. But for oxygen the answer is less clear. The problem has still not been solved. Does the isotopic variation of oxygen have a cause analogous to that which produces the deuterium/hydrogen variation? Variations in the isotopic compositions of the light elements, which are very large and very widespread, seem to be due to the fine chemistry of the molecules in the cosmos; variations in the heavy elements, which are small and often specific to the Allende meteorite, must result from the phenomena of explosive nucleosynthesis, and that we do not yet understand clearly. A new scenario for the formation of the solar system is thus in prospect: an interstellar cloud with interstellar molecules of new products from supernovae injected into it.

Sometimes happy coincidences occur in the development of the sciences. While the cosmochemists were discovering isotopic anomalies in meteorites, astronomers were studying star formation, using, in particular, observations in the infrared and in extremely low frequency waves. These very recent studies have turned out to be extraordinarily fertile and complementary to what I have been saying here. They suggest that young stars are born from *cold and diluted* molecular clouds.

Stars are not generally born one at a time, but in swarms. These swarms disperse very rapidly, and each star takes its distinct place in the cosmos. Groups of stars being born can be observed today in the Orion nebula (Reeves, 1982). It has been verified that the interstellar clouds of gestating stars are illuminated by large, brilliant stars called O and B, which have a very short lifespan (a few million years), and which, after having passed through all the stages already described, explode into supernovae—such at least is the expected scenario, based on the observation of their evolution. If the solar system formed in such an environment, we can well understand the presence of isotopic anomalies and the persistence of aluminum-26. (Recent astronomical measurements have detected aluminum-26 directly in the cosmos.)

Other observations, notably those made by X-ray astronomy, show an intense emission of particles in the molecular cloud region of the Orion nebula, which supports the idea that high nucleosynthetic activity is associated with star formation.

As their name indicates, these cold molecular clouds contain solid particles, within which are organic molecules, the famous interstellar molecules. The organic molecules range from the simple cyanhydric acid to molecules containing alcohol. They have been detected for the last fifteen years by the intensive use of radioastronomy.

Let us see how we can arrange all the available information to sketch out a new scenario for the formation of the solar system. The hypothesis that it was formed by the hot nebula and the condensation of minerals from a gas of solar composition has been disproved. Planetary objects, including the Earth, formed from a cloud of gas and dust, a cold cloud of multiple origins. The gas and dust accumulated gradually over the course of the history of the cosmos. From the Big Bang to 4.5 billion years ago the dust that would become the solid part of the planets condensed in the envelopes of the red giants, the novae or supernovae, or in interstellar space. Slightly before the formation of the solar system, a supernova may have exploded in the vicinity. In the atmosphere of this still-hot supernova, solid grains condensed, producing the Allende inclusions but also a certain number of solid grains of the future planets. These grains and a little newly formed gas were injected into the cold interstellar cloud. This cloud of cold, calm gas, dustier than we had imagined, is the origin of the solar system.

It may seem strange that the explosion of a supernova—a rather rare event in the cosmos (the last explosions observed from Earth were seen by Chinese astronomers in 1054, Tycho Brahe in 1572, Kepler in 1604, and more recently in the Magellanic Clouds—was produced just before the formation of the solar system to scatter its new "seed" into it. This may not be simple coincidence. Some astronomers think that it is *because* there was a supernova explosion that the solar system was able to form. The protosolar cloud's dimensions were too small for it to collapse or implode through its own weight; an exterior impulse was necessary. This impulse could have been the shock wave created by the explosion of the supernova, which, by increasing the density of the cloud, would have helped it to condense. The supernova and the formation of the solar system would therefore simply be two stages in the scenario—a new scenario that could turn out to be just as problematical as the old one.

On the other hand, if the hot nebula hypothesis is dead, did the condensation of grains occur? The arguments developed by Grossman and others for condensation are still valid. Where did such condensation occur? If it didn't occur in the solar nebula, we have to admit that it occurred before that. In interstellar space? In the supernova envelope? In the neighboring red giants? Condensation has to be completely separated from accretion. These two phenomena are distinct and separated by millions of years. This kills the heterogeneous accretion scenario—almost.

But if it is so, why is the volatile character of an element so important in explaining the difference in the composition of the different planets? Should we invoke a reheating created by the early sun?

The fact that this scenario is based on the ultraprecise isotopic analysis of a few grams of rock may be the most beautiful testimony yet to the chronicles recorded in rocks, or rather, to the atoms in rocks, because by transcending the superficial writing that expressed itself in the structure of minerals, the meteoritic palimpsest has revealed its true message to us.

8

Societies of Atoms

The Earth is a product of the cosmos. The chemical elements that form the Earth were among those that were synthesized in the cosmos from the time of the Big Bang to the explosion of the presolar supernova. What is the relationship between the chemical composition of the universe and that of the Earth, between the cosmic abundances of the elements and their concentrations on Earth?

All the elements known in the cosmos exist on Earth, but they are not present in the same proportions. Thus hydrogen, which is by far the most abundant element in the cosmos, is not very abundant on Earth. In contrast, iron, which is important but not very abundant in the universe, is one of the most abundant elements on Earth. Helium, number two in the cosmos, is a minor component of the Earth, much less important than oxygen or silicon. The hierarchy of the abundance of elements on Earth is different from that in the universe.

But this abundance, this hierarchy, is not uniform for the whole Earth. The Earth consists of various structures, various reservoirs, and each reservoir has its own composition and its own chemical signature. The atmosphere is rich in nitrogen and oxygen, the ocean is almost exclusively hydrogen and oxygen combined to form water; the continental crust is rich in silicon and oxygen, but also in aluminum. The mantle too is rich in silicon and oxygen, but also in magnesium. The core is the central region where iron and nickel reign. Thus, not only are the abundances of the elements on Earth different from those in the cosmos, but each part of the Earth, each domain, has its particular chemical composition.

Why does Earth's atmosphere not consist of carbon dioxide like those of Mars or Venus, or of helium and hydrogen like that of Jupiter? Why has the continental crust preferentially concentrated aluminum? In answering these questions we will discuss how a certain reservoir or a certain terrestrial rock formed, why it attracted one particular element rather than another. On the one hand we are responding to a curiosity about chemical composition, on the other to a geological question. Earth chemistry—geochemistry—is an extension of geology, a microscopic approach to geology. Geochemistry is to geology what atomic physics is to classical physics. To say that the continents are a concentration of silicate of aluminum and the mantle of silicate of magnesium is to explain the manner in which the continental crust is differentiated from the mantle. To say that during alteration iron and aluminum remain in place in the ground while sodium and calcium are carried away by flowing water is to describe the process that erodes the continents, and so on (see Figure 47).

But to form this picture must we study the geological behavior of each of the ninety-two chemical elements in the cosmos one by one? This entire book would not be sufficient for such a catalogue, which would be very dry and tedious. Fortunately, we can greatly simplify the procedure because of two fundamental characteristics of the behavior of the elements on Earth: their relative *abundances* and the related properties that allow them to be grouped in *families,* so that we can study the behavior of families rather than of individuals.

The cosmic abundance of an element is determined by the complexity of its nuclear structure. The terrestrial abundance of an element is governed by the nature and the structure of the electrons that surround its nucleus. They determine how one atom can bond with another. The external electrons determine the compounds that elements can form, their nature and structure, and therefore all their chemical properties, artificial or natural. The external electrons are in a sense the "arms" of the atoms, the organs that allow the atom to associate, to be "sociable," to participate in groupings that are called molecules and crystals. In the cosmos we were in the realm of *nuclear chemistry;* on Earth we encounter the *chemistry of electrons* (see Appendix).

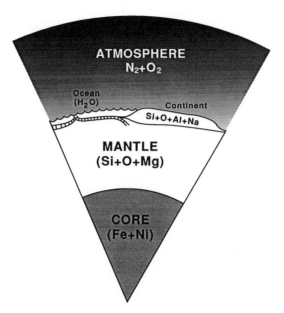

Figure 47 A section of the Earth showing the distribution of the major elements among the different reservoirs.

Mendeleyev's Periodic Table

To anyone interested in simple logic, the chemical elements form a succession, a group of objects classed from 1 to 92. These numbers indicate the number of electrons (and protons) that the atoms of the successive elements contain, starting with hydrogen (1) and ending with uranium (92). They are called atomic numbers, and they are essentially the registration numbers of the atoms.

For the chemist who knows the 92 elements through their behavior, the compounds and molecules they can form, and their properties (to be soluble or insoluble in such and such a solvent, to be more or less volatile, and so on), the reality is completely different and much richer. Each chemical element has its specific characteristics, its own behavior, its specific personality, its affinities and forbidden liaisons: chlorine, for example, bonds well with sodium but never with fluoride; magnesium bonds well with silicon and oxygen but has great difficulty bonding with carbon. Matching up these characteristics to find groupings or simple relationships, we realize as we proceed through the elements from 1 to 92 that there is a marked *periodicity* of behavior. This periodicity appears clearly no matter what quantitative index we use as a function of the atomic number of the element, for example, the ease with which an atom can lose an electron (called *ionization potential*). Far from changing progressively from the lightest to the heaviest either by increasing or by decreasing, the different parameters vary according to a certain periodicity. Thus the elements with atomic numbers 3 (lithium or Li), 11 (sodium, Na), 19 (potassium, K), 37 (rubidium, Rb), and 55 (cesium, Cs) have similar chemical and physical properties and form the alkaline family; the elements with numbers 9 (fluoride, F), 17 (chlorine, Cl), 35 (bromine, Br), and 53 (iodine, I) form the halogene family; and the elements with numbers 2 (helium, He), 10 (neon, Ne), 18 (argon, Ar), 36 (krypton, Kr), and 54 (xenon, Xe) make up the tribe of rare atmospheric gases discovered by William Ramsay, whose common characteristic is to be inert, to bond with no other elements, and to form no compounds. In the three families chosen, we should note that we go from the atomic number of an element to that of its brother by adding 8, and for the last two, 18. These numbers are important. In contrast, neighboring atomic numbers do not imply chemical relatedness. Thus, 19 (potassium, K) has chemical properties nearer to those of 37 (rubidium, Rb) and 55 (cesium, Cs) than to those of the two elements that adjoin it, 18 (argon, Ar) and 20 (calcium, Ca).

In systematizing the periodic character of the behavior and the properties of the chemical elements in 1869, the Russian chemist Dmitri Mendeleyev arranged them in a table called the *periodic table,* which consists of eight columns subdivided near the bottom into eighteen columns. In this table there are two types of associations or groupings:

- Vertical families defined by columns (those we have just mentioned).
- Horizontal affinities defined by proximity.

The *vertical groups* are pronounced at and almost exclusive to the top of the table (elements with low atomic numbers), but the *horizontal affinities* become more important as the atomic number increases. At the bottom of the table there are thus actual horizontal associations of elements with similar chemical properties. Copper (Cu), silver (Ag), and gold (Au), are certainly in the same vertical family, which does not prevent copper from resembling in some ways its neighbor zinc (Zn), or silver (Ag) from resembling its neighbor cadmium (Cd) in its laboratory behavior as well as in its natural associations. For some, the horizontal affinities are so great that they form natural families: this is the rather special case of the rare earths, but also of the iron-cobalt-nickel (Fe-Co-Ni) and osmium-iridium-platinum (Os-Ir-Pt) trios. They are found together in natural deposits as well.

The analysis set forth with such patience and ingenuity by Mendeleyev using data on the properties of the elements was explained when the mysteries of atomic structure were penetrated. The peripheral electrons that gravitate around the nucleus are not distributed randomly but occupy successive orbits. The orbits become saturated when they contain 2, then 8, then 18 electrons, which explains the periodicities in the table.

The Relative Abundances of the Major and Minor Elements

In nature the chemical properties of the elements with their periodic character, their family and tribal groupings, are accounted for by an additional piece of data, their abundance. In the cosmos the light elements with simple structures and low atomic numbers at the top of Mendeleyev's table are the most abundant. Is this also true for the Earth?

To a first approximation we can describe the chemical composition of the great envelopes of the Earth with only *twelve elements*. These are hydrogen (H), carbon (C), nitrogen (N), and oxygen (O) on the first and second lines of the table; sodium (Na), magnesium (Mg), aluminum (Ae), silicon (Si), and sulfur (S) on the third line; potassium (K) and calcium (Ca) at the beginning of the fourth line; and iron (Fe) and nickel (Ni) at the end of it. All these elements appear high on the periodic table, except for iron (Fe) and nickel (Ni), and they are certainly quite high in the abundance curve of the cosmos. Within each terrestrial envelope, however, their abundances are extremely variable: high here, low there. Four or five elements dominate each envelope: silicon (Si), oxygen (O), and magnesium (Mg) dominate the mantle; iron (Fe) and nickel (Ni) constitute the core; nitrogen (N) and oxygen (O) are the major components of the atmosphere, and so on. The eighty other elements in the table, the vast majority, represent less than 1 percent of the mass.

The abundance of the chemical elements on Earth does not vary progressively. There are the *majors,* and the rest are the *minors,* whose abundance is not measured in percentages but in parts per million (ppm) or parts per billion (ppb). The minor elements do not play an essential role in forming the chemical compounds that dominate the planet although they include elements as important as uranium (U), silver (Ag), gold (Au) and platinum (Pt).

If the relative abundance of the elements is an essential fact, the notion of the chemical family or tribe is equally important for geology. On Earth the minor elements associate with the major element in the same family or, by default, with the one they resemble the most and which is in a way their guide: for example, rubidium (Rb) (minor) is faithfully associated with potassium (K) (major); cobalt (Co) with iron (Fe), gallium (Ga) with aluminum (Ae), germanium (Ge) with silicon (Si), and selenium (Se) with sulfur (S).

Although low in abundance, the minor elements are not negligible. Some, like uranium (U) and thorium (Th), are radioactive and the most important source of internal energy for terrestrial geologic phenomena. Others, like silver (Ag) or gold (Au), or more modestly, copper (Cu) or molybdenum (Mo), play a major economic role. Finally we shall see that using the notion of the family helps in tracing the evolution of the great geologic phenomena. In this game the minors play a major role in decoding geologic messages because for each reservoir their abundances vary, while the abundances of the majors are quite constant.

Atoms, Minerals, and Rocks

Atoms bond with each other to produce chemical compounds. Two atoms of hydrogen and one of oxygen bond to produce a molecule of water, the essential geologic agent of the terrestrial surface. The bonds of two atoms of nitrogen or oxygen form the two essential molecules in the atmosphere. Three atoms of oxygen bond to form the very important compound ozone in the upper atmosphere. This is all familiar elementary chemistry.

In the solid Earth the chemical compounds that interest us are more complex. They consist of giant molecules made not of one, two, or ten atoms, but of thousands of thousands of atoms. These compounds are then natural crystals called minerals. Minerals are numerous and varied, and studying them is a discipline in itself (mineralogy). Although oxides, sulfides, and carbonates are important, if we want to simplify and go directly to the essential, one family of minerals, the *silicates,* is more important than all the others. As a gross simplification we could say that geochemistry is in large part the chemistry of silicates (see Figure 48).

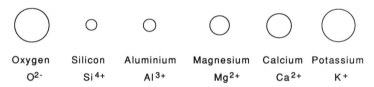

Relative size of the principal ions in silicate formations

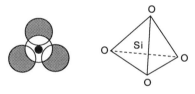

The silicon-oxygen tetrahedron [SiO_4]

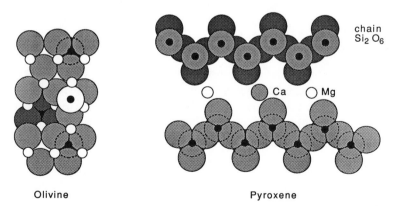

Figure 48 The elementary structures of silicates from the ions through the elementary silicate tetrahedron to the two most important silicates of the upper mantle, olivine and orthopyroxene.

Silicates consist of two parts: a framework or *matrix,* and *tenants.* The matrix is a vast network formed by a basic tetrahedral structure, the SiO_4 tetrahedon. The tenants are a whole series of ions such as aluminum, sodium, magnesium, potassium, and calcium. The matrix and the tenants interact according to the size of their ions, their charges, and temperature and pressure conditions. The great variety of the family of silicate minerals, with its preferential associations of atoms, is created in this way.

The silicates that are found in the crust near the surface have an open structure that can hold large atoms like potassium and aluminum, and large holes. Mantle silicates that are often subjected to high pressure are compact and dense and hold only atoms of small dimensions, such as magnesium. In fact, the mantle is almost a pure silicate of magnesium. Looking at a silicate's structure is a way of finding out what medium it originated from.

In nature minerals are not generally isolated. They mix together in those complex but familiar assemblages: rocks. Rocks are mixtures of minerals. Rarely are they insignificant mixtures based on chance encounters; more often they are associations that obey precise rules. Certain minerals such as quartz and olivine cannot cohabitate; they react in each other's presence to create a new mineral (pyroxene). Others have such an affinity that one is rarely encountered without the other: olivine and pyroxene, plagioclase and pyroxene, and potassium feldspar and quartz are the famous couples in the rock science called petrology. In fact, most mineral associations are in thermodynamic equilibrium and obey precise laws.

As minerals are societies of atoms, rocks are societies of minerals. *Society* is defined as association by rule, in contrast with *mixture,* which is a random association. In other words, a rock is an association of various minerals but not a random association. The architecture of the terrestrial crust is clearly divided into levels of organization: atoms, minerals, and rocks. The rocks form the fundamental structural units of the Earth, continental or oceanic, crust or mantle.

The Earth, a Chemical Factory

The Earth consists of varied chemical compounds, but its composition was not defined once and for all, unchangeable and immobile. The various natural chemical compounds, molecules, and crystals in

Paul Gast of the Lamont-Doherty Geological
Observatory was one of the founders of
modern geochemistry. He died at the age of
forty-three, at the height of his scientific
achievement.

the solid, liquid, or gaseous state have come apart, reassembled, combined, and reacted with each other to produce other chemical combinations throughout geologic history. Natural chemical reactions are
processes through which natural compounds are produced and
transformed into other compounds. These transformations combine
to produce the great geologic phenomena. The geological processes
of erosion, sediment formation, generation of magmas, and generation of mountains are all combinations of chemical transformations.
The Earth is thus an immense chemical factory that constantly produces, destroys, transports, recombines, dissolves, and precipitates
tons of chemical compounds in the oceans, on the continents, and in
the Earth's interior, right down to the core (see Figure 49).

Like all chemical reactions, those that take place in nature involve
breaking the chemical bonds of some compounds to form others. But
in nature these transformations would play no role if they were not
amplified by the transport of materials, which separates, sorts, and

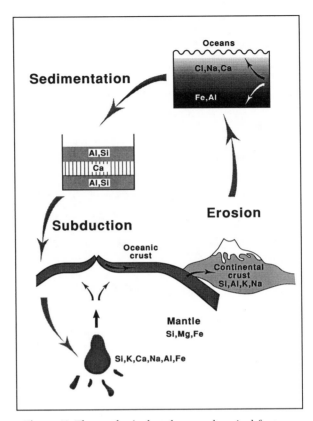

Figure 49 The geological cycle as a chemical factory.

classifies the different products. The reactions that take place during surface alteration and erosion destroy the silicate minerals of internal origin. Some soluble ions are freed and transported invisibly with fresh water toward the sea. Other compounds remain in place in the ground or are evacuated as particulates. Sodium and calcium are evacuated with water. Iron and aluminum stay in place creating soils (and in the extreme case the so-called laterite). A gigantic chemical sorting takes place on the Earth's surface and water is its principal agent.

Convective currents occur in the interior of the Earth, and the resulting pressure release destroys certain chemical bonds and causes some minerals to melt. This phenomenon is followed by the transport of the lighter parts of the melt toward the surface and, with them, the particular elements that it has concentrated. In this way potassium, aluminum, and calcium are expelled from the mantle and transferred to the surface in volcanic magmas. Chemical transformation and transport are separate processes whose geologic roles complete and reinforce each other to effect continuous change in the chemistry of the planet.

The chemical operations whose theater is the globe and whose mechanisms are sometimes surface phenomena, such as alteration, transport, and sedimentation, and sometimes interior phenomena, such as magmatism and metamorphism, can be divided into two types:

1. Those that separate, isolate, or differentiate some elements from others to produce new chemical entities and structures. Examples are the concentration of calcium carbonate in the shells of aquatic animals, the separation of calcium from aluminum by erosion, and the separation of magnesium from potassium by magmatism. All of these processes that differentiate elements to create new structures (limestone in the first instance, soils in the second, lava flows in the last) are called *differentiation processes.*
2. Those that mix, homologize, and destroy organized structures. Such is the case in mechanical erosion and detritic sedimentation, and in the convective movements of the mantle that tend to mix all of the different structures to produce a more or less homogeneous medium. These are called *mixing phenomena.*

The first concentrate certain elements; the second disperse them. The first produce order; the second simply organize the disorder. We will look at a clear illustration of this duality, which dominates all geological phenomena.

Minor Elements and Ore Deposits

The low-abundance minor elements are clearly not the components of the principal minerals in the terrestrial crust. There they are found only as impurities or disguised inclusions. They play no role in geochemical reactions, content to distribute themselves as well as they can among the products of the reaction. In this they follow the major element they most closely resemble.

They are, however, constantly used by man as in some cases they have been for millennia. This is the case for gold, silver, and platinum and also, more modestly, for lead, copper, and tin. Some are essential to modern society, uranium certainly, but also the rare earths, which are used in television screens, and chromium and titanium, which are used in the alloys from which spacecraft are made. How do we get them? Do we have to pulverize cubic kilometers of rock to recover the few tons we need?

Many of the minor elements have a strange and fascinating property: in spite of their low abundance they have managed to concentrate themselves in certain places, where they form ores, well-defined compounds, and visible crystals that are sometimes massive and ornamental. This is the case for copper, lead, zinc, and molybdenum, which form alloys with sulfur to produce sulfides with a brilliant luster (chromite, titanite, titanomagnetite, stibnite), and for chromium, titanium and tin, which bond with oxygen to produce oxides. In veins, lodes, or beds the metallic deposits are like anomalies of nature, abnormal and localized concentrations of minor elements. In a chromium deposit, that element is three thousand times more concentrated than in crustal rock; tin in a mine is twenty-five hundred times more concentrated and lead four thousand times, but nickel only fifty times more than in peridotite. But this virtue is not universal. Some minor elements do not concentrate well. If they are to be useful they must be extracted from enormous quantities of sterile rock. This is the case for minor elements such as gallium and germanium, and to a lesser extent, the rare earths. All these properties have an economic consideration: their cost, which certainly depends on demand but also on how difficult it is to obtain them. Gold is ten thousand times more abundant than copper yet costs ten thousand times as much. Gallium, which is forty thousand times more abundant than gold, costs only five times less because it concentrates much less well.

The Geologic Tribes of Elements

We are beginning to understand how the ninety-two chemical elements behave on Earth. Combinations of the major elements explain the composition of the principal terrestrial reservoirs: atmosphere, hydrosphere, crust. The minor elements are distributed among these reservoirs, richer in some than in others but always unobtrusive. Sometimes these minor elements form exceptional concentrations that man exploits in mineral deposits or mines.

We know that there are linkages and correlations among all these distributions suggested by the layout of Mendeleyev's table. But can we learn more from geologic groupings based on the chemical similarities among elements? One of the fathers of Earth chemistry, Victor Goldschmidt of Germany, approached this problem nearly forty years ago (1954). Using analyses of terrestrial rocks, the seas, and meteorites, he regrouped the elements of Mendeleyev's table into four classes or geologic families:

- *Atmophiles:* the elements of the atmosphere and the hydrosphere, which, besides nitrogen, oxygen, and hydrogen, include the rare gases: helium, neon, argon, krypton, and xenon.
- *Lithophiles:* the "stone lovers," which are preferentially located in the silicates: potassium, calcium, sodium, magnesium, silicon, and aluminum (the elements that form the silicates). The minor elements that resemble them are also included. Among those that resemble potassium and are members of its chemical family are rubidium and cesium; those that resemble calcium, strontium and barium; those that resemble silicon, germanium; and those that resemble aluminum, gallium, but also other elements that have less direct connection with the major elements, like the rare earths, uranium, and thorium.
- *Siderophiles:* the "iron lovers," which include, besides iron itself, nickel, cobalt, osmium, rhenium, iridium, and gold. They are abundant in iron meteorites and probably in the core of the Earth.
- *Chalcophiles:* the "sulfur lovers," which are concentrated in exploitable mineral deposits in the form of sulfides. They include copper, lead, zinc, and arsenic.

Goldschmidt's families consist of a few major elements, each of which is, as it were, the head of a family, and an abundant cohort of minor elements whose chemical properties resemble those of their head. Each geologic tribe defines a zone or domain in Mendeleyev's table. For Goldschmidt the distribution of the geologic families in the various terrestrial reservoirs was simple: the atmophiles in the atmosphere, the siderophiles with iron in the core; the lithophiles in the crust and the upper mantle; and the chalcophiles in the deep mantle. The degassing of sulfurous fumes by volcanoes and the famous sulfur odor associated with eruptions were considered an obvious proof of the abundance of sulfur in the mantle.

By arranging the chemical families among the various terrestrial envelopes in this way, Goldschmidt accomplished a scientific feat of considerable importance: he established a direct relationship between the distribution of the elements in Mendeleyev's table and their distribution in the terrestrial envelopes (see Figure 50). He affirmed the existence of a direct link between the structure of atoms and that of the universe as a whole. He realized that enormous features in fact resulted from the intimate structure of the matter of which they were the faithful transcription. This is a transfer on a scale that erases all the intermediary steps—minerals, rocks, and rock for-

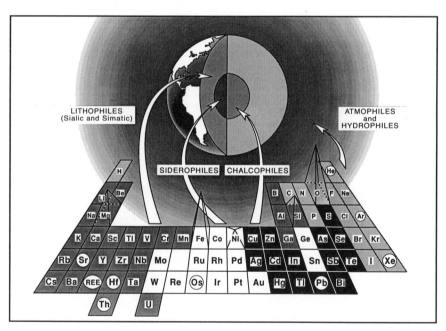

Figure 50 Mapping Mendeleyev's periodic table to the structure of the Earth (the different geochemical families are in different grays).

when the piece of the crust from which it came was segregated (see Figure 51).

No chemical analysis would have allowed us to draw such conclusions, since when the rock was formed it was totally changed by the process that produced it. If it was a magmatic rock, fusion totally changed its composition, yet its isotopic composition remained the same. The isotopic composition of the source was transcribed into that of the magma. The isotopic message traversed the vicissitudes of geologic life but kept its memory of previous events.

Since volcanism is the most widespread geologic phenomenon and brings magmas to the surface from depths of hundreds of meters, the isotopic analysis of volcanic lavas is a veritable chemical messenger of the Earth's interior and its past: of the interior because volcanoes

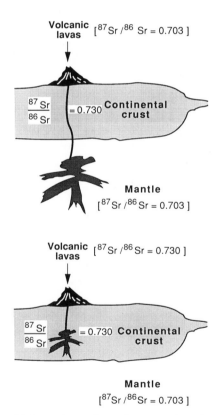

Figure 51 Isotopic tracing using strontium isotopes.

have their ultimate source several hundred kilometers under our feet in the mantle; of the past because for the isotopic message to take shape it needs a certain amount of time to form impressions. Only those events old enough to have changed the rubidium/strontium ratio have been reflected in the isotopic message. It is a spatial and temporal tool of formidable usefulness.

This example is extremely simplified. It does make it clear, however, that the making of a set of isotopic data for the principal terrestrial reservoir and of mathematical simulation models has given us a very powerful method of investigation. The isotopic compositions of the heavy elements are tracers of geologic phenomena, just as short-period radioactive phenomena are tracers of biologic phenomena.

The rubidium/strontium example is not unique. All the geochemical families have their isotopic tracers. The atmophiles have argon, helium, and xenon, whose isotopic compositions vary as a consequence of the radioactivity of potassium and uranium. As we will see, they record degassing phenomena. The lithophiles have many geologic tracers. Besides strontium-87, which we have already described, there is neodymium-143 (a product of samarium-147) and hafnium (a product of lutetium-176). The siderophiles have osmium-187 (a product of rhenium-187), and the chalcophiles have lead, whose three isotopes vary (as we have already seen in Patterson's work on the age of the Earth). These tracers will help us to decipher the essential phenomena of the chemical (and therefore geological) evolution of our planet.

From Sun to Earth

The elements are distributed in the various terrestrial reservoirs according to Goldschmidt's grouping by families. To get a global view of the composition of the planet we have to make an inventory of all these reservoirs. Then we can try to understand the nature of the phenomena that start with cosmic abundances and end with the composition of the Earth, and in this way retrace the journey from star to stone.

This inventory is difficult to make. Not only does the Earth consist of several reservoirs, but they are themselves heterogeneous, and some of them are difficult to sample. What do a granite and a limestone have in common? A basalt and a schist? We have some rocks from the upper mantle that have been brought to the surface by volcanoes, but none from the lower mantle, and certainly none from the core. All we have for the deep layers are indirect indications provided by seismology. How can we hope to calculate a valid average chemical composition for the Earth from such disparate information? One of the great successes of geochemistry is that we have in fact succeeded in making quite a precise estimate. I will simplify the steps involved in order to follow the central reasoning process.

The basic reservoir is the mantle. From it the continental crust on one side and the core on the other separated out and became autonomous. Oceanic crust is continuously extracted from the mantle, but continuously rejected as a result of seafloor spreading, so it does not play a role. The present mantle is therefore a residue. Its composition can be approached in two ways: on the one hand, the basalts emitted by volcanoes are products of the mantle; on the other, volcanoes bring pieces of periodotites to the surface as exotic fragments that are untransformed pieces of the mantle. By performing a series of multiple cross-checks and making extensive use of the information provided by the isotopic composition of geologic tracers, we can reconstitute the chemical composition of the mantle for both major elements and large numbers of trace elements. Since it is subject to large internal movements reflected at the surface in continental drift, one can assume that it is well "mixed," and that the chemical information obtained about the upper mantle can thus be generalized to the mantle as a whole.

The crust is more directly accessible. On the continents geologic cartography gives each rock its relative importance; dredging and drilling have shown that basalt is the major component of the ocean floor.

The situation is more difficult for the core, because we certainly have no direct access to the center of the Earth. Since the 1960s we have known that it consists of an iron-nickel alloy. We also know, thanks to the chemical analysis of meteorites and to laboratory experiments, how the elements partition themselves between iron-metal and silicate. We determine the ratios of the concentration of the element in iron and in silicates when they are in contact. For example, for osmium the ratio is 10,000 in favor of iron, for lithium it is 0.02, which means that lithium goes into the silicate. Knowing the composition of the silicate mantle, we can calculate that of the core. If the concentration of an element in iron is fifty times its concentration in silicate, this measurement of its concentration in the mantle makes it possible to compute its concentration in the core.

Using this reasoning, which I have simplified outrageously, we arrive at an estimate of the chemical composition of the various envelopes and can make a weighted average to obtain an average composition for the Earth. A number of scientists have completed this exercise in recent years using various approaches, and we are now converging on a composition agreed upon by everyone to an accuracy of a few percent (Ganapathy and Anders, 1974; Wänke, 1983).

Now we can compare the average composition of the Earth with that of the sun, which is a necessary reference point for any object in the solar system. We realize that the Earth is enriched in some elements and depleted in others, which means the same thing, since the total of a chemical analysis is always one hundred percent, but the amounts of enrichment and depletion do not correspond to Goldschmidt's families.

Some lithophiles are enriched, others depleted. All the atmophiles are depleted, but some more than others. The chalcophiles, beginning with sulfur, seem very depleted. What logic determines the Earth's chemical composition? Information obtained from studies of meteorites can help us here. Anders's work showed the importance of the relative volatility of an element, and we have set up a scale of elements from the refractory to the volatile.

We can apply these ideas to the chemical composition of the Earth as compared to that of the sun (see Figure 52). Terrestrial rocks are depleted in volatile elements, as are metamorphosed chondrites and lunar rocks. The degree of volatility affects the major elements but also the minor ones. Intensive use of all the elements then allows us to measure the exact amount of loss of the volatiles (or more exactly, the nonincorporation of the volatiles) on Earth. This illustrates the necessity of studying *all* the elements, not just the major ones, because they offer a larger palette or broader range for comparative chemistry. The depletion in volatiles is illustrated by the strontium isotopic ratios of the mantle. The present-day Earth mantle has a strontium-87/strontium-86 ratio of only 0.703, while chondrites have ratios of 0.740 to 0.750 for similar strontium contents. This shows that the Earth has lost its rubidium, the radioactive father of strontium-87. Rubidium is a volatile element.

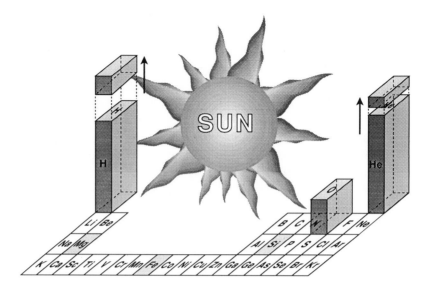

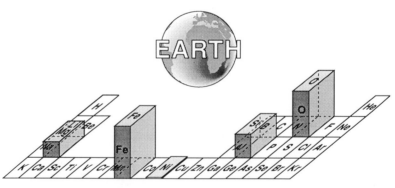

Figure 52 Chemical abundance in the sun and in the Earth. The height of the blocks indicates relative abundance.

Goldschmidt's family scheme does not turn out to be very useful in determining the global composition of the Earth based on that of the sun. On the other hand, Mendeleyev's table and atomic structure are fruitful and reflect equally the degrees of volatility. Arranging all the major and minor elements according to volatility results in groupings on the periodic table that differ from Goldschmidt's. If we want to construct a *universal geochemical-cosmochemical classification* we must complete Goldschmidt's classification with a volatility index.

We also have to make some modifications to take account of progress made in the last twenty years. In the lithophile family we must today distinguish two subfamilies: that of elements concentrated in the continental crust—the *sialic* elements such as potassium, rubidium, aluminum, and uranium; and that of elements concentrated in the mantle—the *simatic* elements such as magnesium and chromium. They have totally different geologic behaviors. We must also modify Goldschmidt's arrangement in another way: contrary to what he thought, the lower mantle is not rich in sulfur and chalcophiles, but has a lithophile composition just as the primitive mantle may have had. Seismic observations together with high pressure experiments have eliminated the hypothesis that the lower mantle is made of sulfur.

Little by little the classification into distinct and exclusive geochemical families is today giving way to a typology of multiple affiliations. We are preserving Goldschmidt's reasoning but making it richer and more flexible.

History in Two Episodes

The composition of the different terrestrial reservoirs can be explained as the result of two successive gross chemical processes. After terrestrial abundances were determined by cosmic processes, the elements distributed themselves among the various reservoirs. How did these two successive differentiations take place?

The first took place in the cosmos when the Earth was differentiated from a cloud of gas and dust. At the moment the Earth was formed, it lost its volatiles and thus became enriched in refractories. The chemical elements were present in the nebula in cosmic abundances. The condensation and agglomeration of the planet acted like a chemical filtration system to sort the elements and establish a new set of abundances. I will devote some time to the differentiation of the terrestrial envelopes, because it affects our entire understanding of geology. The failure of the hot nebula scenario complicates our attempt to understand cosmic differentiation. In the traditional scenario the separation between volatiles and nonvolatiles was defined by the planetary condensation temperature, the one at which condensation stopped, which in turn depended on the heliocentric distance of the planet. We have substituted the scenario of the cold nebula in which the grains and the dust were born earlier, either in interstellar space or in the envelope of the presolar supernova. Now we have to construct the Earth from a cold cloud of gas and dust.

Let's see how it can work. The inner planets accreted from the dust available in the presolar disk. But the disk was increasingly hot toward its center, so the dust was subjected to intense radiation as it approached. The solid dust consisted of some oxides of aluminum and titanium (the equivalent of the famous Allende inclusions), pure iron, and silicate and sulfide grains. The metallic iron grains were heavier and more resistant to the heat than the silicate and sulfide ones. The iron/silicate ratio was therefore higher toward the center of the disk than near the edges. This explains the high density of the planet Mercury, which is closest to the sun.

The grains agglomerated little by little, like a snowball, including a little interplanetary gas among them. When all the solid balls had formed but had not reached the size of a planet, the central star, a "baby" sun, began to shine brightly, like all stars in the formation process. Its brilliance was far greater than that of the adult sun of today. From it blew a wind of particles and extremely intense radiation that swept all the gases from the inner corona. This sealed the fate of the inner planets: they would not have impressive gaseous atmospheres; that would be reserved for the external planets. Jupiter would be the most voluminous, since it would be located at the edge of the zone swept by the primitive solar wind, where some of the gases blown away by the wind would accumulate. Along with these gases Jupiter may also have inherited some of the sun's angular momentum—perhaps.

We can complete the picture by putting the planetary differentiations into a precise chronological context. According to Goldschmidt's scenario the composition of the Earth as we know it today, with its concentric envelopes of core, mantle, crust, and atmosphere, goes back to the dawn of geologic time when it was created, 4.5 billion years ago. Planetary differentiation was only a consequence of planet formation. It must have taken place very early in Earth's history.

In order to picture what must have happened at that time, Goldschmidt used the analogy of a blast furnace. Iron ore is melted in a furnace without oxygen and transformed into metallic iron. The iron begins as an oxidated ore, gains two or three electrons, passes to the state of metallic iron, and falls to the bottom. The remaining silicates form a light slag and float on the melted bath like a crust. The reaction gases are given off in a heavy steam. According to Goldschmidt, at the dawn of geologic time this is how the Earth, which was almost totally molten, degassed its atmosphere, while the metallic iron fell to its center to form the core. The mantle was formed by heavy sulfides and silicates rich in magnesium. The continental crust was formed by lighter silicates rich in calcium, potassium, and sodium, the "geologic slag" that floated to the surface of this gigantic ocean of magma.

This view of archaic differentiation leads naturally to the idea that Earth's history consisted of two episodes: a rather brief archaic period, which was tumultuous, full of noise and furor, and inaccessible through geologic methods; and a long series of cyclic episodes—including erosion, metamorphism, and orogeny or mountain formation—which were constantly renewed and repeated, which were similar to each other, and whose geology could be studied but whose cumulative effects remained insignificant for the deep structure of the planet. Geology was thus restricted to the "epidermis" and to recent times. Using a religious metaphor we could say that history was divided into a Genesis, the act of a god, whose chemical effects we could only imagine, and a much longer geologic time, which human science was able to understand.

We can contrast this scenario with another more gradual, more evolutionary one. In the latter the various terrestrial reservoirs—continents, atmosphere, core, oceans—were formed progressively and continuously over the entire course of geologic time. Two distinct episodes or a continuous evolution of the Earth over a long and irreversible time period: which is correct?

9

Planet Earth

The idea that Earth's history was divided into two episodes, the beginning and what came afterward (the first belonging to astronomy and the second to geology) was for a long time firmly fixed in people's minds. Until 1960 the interior of the Earth was thought to have cooled off long ago and to be geologically insignificant. The interesting transformations of geologic activity, the ones that had been recorded in the rocks, were supposed to take place on or near the surface, without any influence from the interior. Of course, the interior manifested its existence from time to time in violent paroxysms called earthquakes and volcanic eruptions, but these were only epiphenomena of a resolutely Huttonian geology. If the interior was passive, the position of the oceans and the continents was also supposed to be immutable, having been established once and for all during the original great upheaval. The borders between the continents and the oceans certainly changed over time, but this was only the result of essentially vertical movements (up and down) in which the continents alternately rose or subsided, initiating the vast retreats and advances of the sea that are called marine regressions and transgressions. To surface geologists the mantle more than 50 kilometers beneath their feet seemed very far away, very inaccessible, very useless, very uninteresting.

Around 1960 this fixed view began to collapse. Today we know that earthquakes and volcanoes are not minor events for the planet but vigorous indicators of an intense, powerful, and influential inner life. In fact, they govern a large number of geologic phenomena that, contrary to what Goldschmidt thought, are constantly changing our planet and its surface.

From Continental Drift to Plate Tectonics

Between 1910 and 1929 the German meterologist Alfred Wegener tried to convince the geological and geophysical community of the existence of horizontal movements in the terrestrial crust, that is, of continental drift, but without success. In the 1960s, almost fifty years later, the horizontal movement of the terrestrial surface—continental drift and seafloor mobility—finally reappeared. Without repeating a series of events I have recounted elsewhere (Allègre, 1988), let us retrace the principal features of what is today called the theory of *plate tectonics*.

At the crests of the great linear submarine mountain chains (called oceanic ridges) that crisscross the world's oceans in a network 60,000 kilometers long, ocean floor is constantly being formed. The material is provided by an almost uninterrupted series of volcanic eruptions. The oceanic crust formed in this way drifts off on both sides of the ridges at an average rate of two centimeters per year. The oceanic ridges thus create terrestrial surface. Since the total surface area of the Earth does not increase, the phenomenon of creation must be compensated for somewhere. This takes place at the great oceanic trenches, where the oceanic crust plunges into the mantle, carrying with it part of the sediments that have been deposited on it during its drift across the ocean floor. This mechanism, called seafloor spreading, proposed by Arthur Holmes (1945) of Scotland and then by Harry Hess (1960) of the United States and demonstrated through observations of magnetic anomalies by Lawrence Morley of Canada and Fred Vine and Drummond Matthews of England in 1963, established that the maximum lifetime of the oceanic crust is 200 million years. The oceanic crust is thus eternally young; since it is continually created and continually destroyed, it does not date from Earth's earliest history.

The mechanism of seafloor spreading was systematized in the theory of plate tectonics, according to which the globe is divided into rigid spherical shells called plates. The plates have three kinds of boundaries: ridges, trenches, and a type of fault called a *transform fault.* The plates are created at the ridges and destroyed at the trenches. Between the two they spread out in a rigid way, without deforming. They are few in number, totalling about a dozen, and their thickness, 100 kilometers, is greater than that of the crust. At plate boundaries, large amounts of internal energy are dissipated in the form of earthquakes as well as in the form of volcanism. The map of earthquake locations corresponds exactly to that of the plate boundaries. These phenomena take place at the ridges, of course, but also at the subduction zones, where the plate descending with its sediments melts to produce volcanoes: thus volcanoes that border the Pacific have products that are much richer in silicon and aluminum than the basalt of the ocean floor (see Figure 53).

What have the continents to do with all this? Consisting of relatively light composite materials rich in potassium silicate and pure silicon, they float on the surface of the Earth. Attached to pieces of

Seiya Uyeda of the University of Tokyo, one
of the founders of plate tectonics.

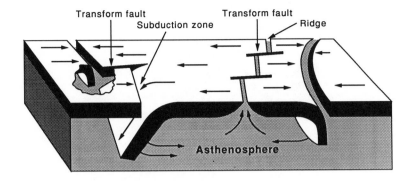

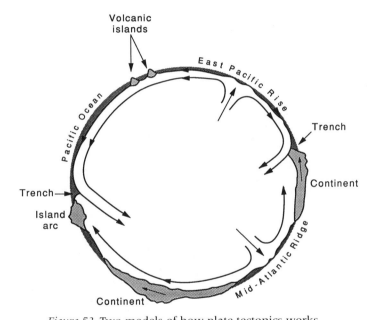

Figure 53 Two models of how plate tectonics works.

oceanic plate, they move with them; they drift but never plunge into the mantle. Unlike the eternally young oceanic floors, the continents seem eternal. Studies since 1970 have shown that their geologic activity is much greater than had been imagined. Sometimes they break up, as the supercontinent Gondwana broke apart about 200 million years ago, producing pieces that are today called South

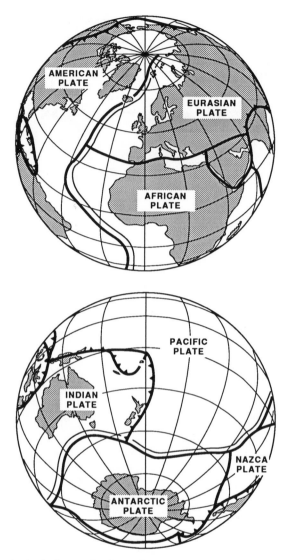

Figure 54 Principal plates. The limits between plates are ridges (double line), subduction zones (jagged line), and transform faults (single line).

America, Africa, India, Australia, and Antarctica. At other times they collide, as when India, which left Africa 120 million years ago and began drifting north, ran into Asia 55 million years ago. This collision produced the Tibet-Himalaya zone and sutured two pieces of continent together to produce a new and much larger one. Thus the continents, light crusts floating on an underlying mantle, break up, drift, collide, adhere, and break apart again. Their movements produce a veritable ballet on the surface of the globe whose rate, a few centimeters per year, is measurable in geologic time (see Figure 54).

But plate tectonics does not even approach the problem of how and when the continents formed. Are they as ancient as Victor Goldschmidt thought? If so, we would have to conclude that continental drift has existed over the entire length of geologic time. Or are they products of the very plate tectonics that emits volcanic lavas with a composition close to that of the continents at the subduction zones? In the latter case, the face of the Earth and the relative size of the continents and the oceans would have constantly varied. This is a fundamental question that involves the entire history of the Earth and all of geology.

The Formation of the Continents

The continents on which we live and which for many are synonymous with the Earth itself, appear at first glance to have extremely varied shapes and compositions. Some regions are flat and covered with sedimentary rock laid down in strata: these are sedimentary basins. Areas of high relief contain rocks folded and broken into mountain ranges. What James Hutton observed and what generations of geologists after him have confirmed is that the granite intrusions in these mountains coincide in time and space with the folding of the mountains. This is the case for the granites of the Sierra Nevada range of California, of Maine and Vermont, of the Appalachians of the eastern United States, and of Makalu, Manaslu, and Everest in the Himalayas.

We can complete this rapid presentation of the continents with a survey of the nature of rocks and the ages of geologic events. Continental rocks vary greatly, from sedimentary rocks such as limestone, sandstone, and schist, to igneous rocks such as granite and basalt. A survey of continental rocks shows that the most abundant of them is granite. If rocks similar to granite are added, they account for more than 80 percent of the continents. Because of their privileged distribution near the surface, sedimentary rocks give a misleading impression of importance. In fact, they form only a thin skin, at most 1 or 2 kilometers of the 35 kilometers of continental crust. Drilling through the horizontal sediments of the basins has brought up samples of folded metamorphosed rock injected with granites. The subbasements of the basins are old mountain ranges that have been eroded and worn nearly flat.

Granite has a very specific composition. It consists of two minerals, quartz and feldspar. From a chemical point of view, it is a silicate assemblage that concentrates silicon, aluminum, and potassium: its aluminum concentration is twenty times greater, and its potassium concentration a thousand times greater than that of mantle rocks. Studies of the geological habitat of the granites show that the majority of granitic rocks are intimately associated with folded rocks.

The cartography of the continental terrains, combined with their dating by radioactive methods, has shown that folded rocks and granites of the same age form regions called *provinces,* whose dimensions are several hundred to several thousand kilometers. Such are the Grenville province in North America, which is a billion years old, and the hundred-million-year-old Alpine province in Europe. Provinces are coupled together, creating a veritable mosaic on the map of the continents. In certain places, such as North America, the assemblage also has a polarity. Around a central nucleus 2.7 billion years old, belts seem to have been molded one after another, becoming increasingly younger toward the edges (see Figure 55) (Windley, 1977).

Gathering all this information together, we can sketch the following scenario: The continents, or rather pieces of continental crust, were created during well-defined geologic episodes, the same ones that raised mountain ranges. Orogenesis, the formation of mountains, is therefore the fundamental process of continental genetics. After the formation of the mountains, erosion acts on the peaks and wears them away until it makes near-plains or peneplains, that is, evolved, adult continents, of them.

Since mountain ranges from 2.7 billion to 30 million years old exist, it is thought that the continents formed progressively over the

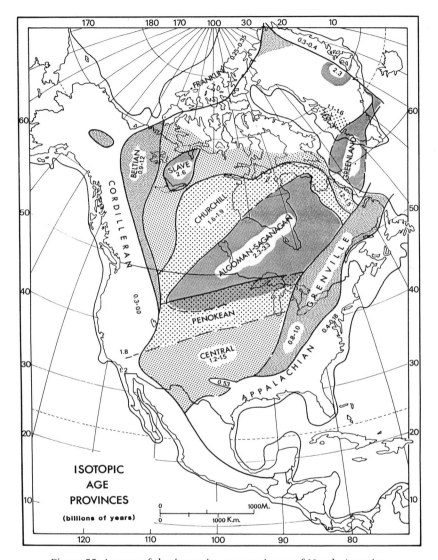

Figure 55 A map of the isotopic age provinces of North America.

course of geologic time. In the "beginning" there were only a few nuclei, which, like waterlilies on the surface of a lake, progressively increased their surface area. Translating the cartography of the continental provinces into histograms we see that the continents apparently grew faster and faster as time passed (see Figure 56). This sug-

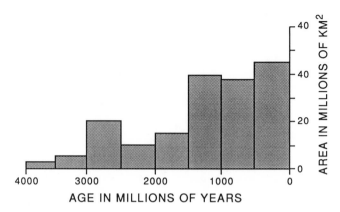

Figure 56 Histogram showing the proportions of continental surfaces from different ages.

gests the question: Are the oceans destined to disappear eventually, or, on the contrary, to submerge the continents?

The above scenario, which seems to account for the essential geologic observations, changes radically if we assume that orogenesis does not create new pieces of continents but only reuses and rejuvenates old pieces of preexisting continents, if, in short, orogenesis creates the new from the old. Let us imagine that all the continents were formed 4.5 billion years ago, but that in each geologic period parts of them were destroyed by erosion and transported to the ocean floor in the form of sediments, which were folded, baked, melted, and transformed into granites during orogenesis and then added to continental crust again. That would be a true recycling of the continental crust. According to this scenario the crust would not have increased in volume over geologic time, it would only have changed its appearance and its geologic age; in places it would have been progressively rejuvenated (see Figure 57).

How can we decide which is correct, the theory that each province contains a majority of new materials recently extracted from the mantle or the theory that the materials had already been crustal for a long time? In order to do so we would have to be able to date when a chemical element, an atom contained in a province, left the mantle to enter the crust. We would have to be able to reconstruct the history of the atoms in the orogenic provinces. Classical geology had no techniques for solving this problem.

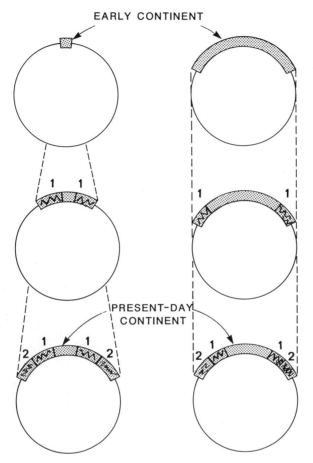

EARLY CONTINENT

PRESENT-DAY CONTINENT

Figure 57 The two competing theories of continental growth: left, continuous creation; right, early creation with continuous reworking.

The Geologic Itinerary of the Continental Atoms

Suppose that in a fairy-tale country everyone received the same amount of money every day. In walking around the country it would be easy to recognize recent immigrants: they would be poor. In contrast, the original habitants, who had been established for generations, would be very rich. The situation of some atoms in the geologic world is very similar. Strontium atoms in the continental crust regularly become richer in strontium-87, which is produced through the radioactive disintegration of rubidium. Since rubidium is a sialic element, the crust is rich in it and the production of strontium-87 is high there. On the other hand, in the mantle, which is poor in rubidium, the amount of strontium-87 produced is small. If we analyze the strontium in a continental rock, its strontium-87 content immediately reveals whether the rock is of ancient continental stock or new, an "immigrant." The case of strontium is not unique. We have at our disposal other lithophile tracer isotopes such as neodymium and lead. We can study a problem by using several isotopic tracers to confirm the results of the first analysis.

Isotopic studies of continental origins begun in 1960 at first gave disappointing and contradictory results. Patrick Hurley of MIT, using

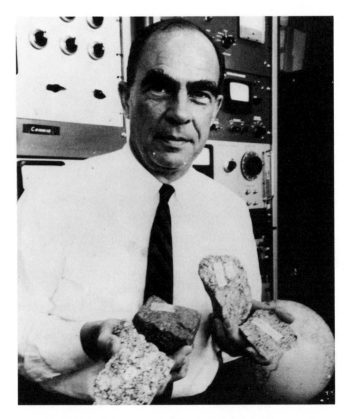

Patrick Hurley of MIT was the first to apply modern geology based on age determinations to the problem of continental drift and continental growth.

strontium isotopes as a tracer, concluded that the continental crust is formed continuously from new material that originates in the mantle (Hurley et al., 1962). Clair Patterson (1963) of Caltech, who used a lead isotope as a tracer, arrived at the opposite conclusion: that the continental crust differentiated very early in Earth's history (3.5 or 4 billion years ago?), and that since then, orogenesis has simply recycled the same continental material.

Only when the neodymium tracer method was perfected did it become possible to reconcile these points of view to reach a surprising conclusion: in each piece of continent dated geologically, part is ancient and recycled but part is new and newly formed (Allègre and Ben Othman, 1982). In brief, the continents are neither old nor new but a mixture of the new and the secondhand. Contrary to Goldschmidt's view, most of the continents did not form in the primitive period of Earth's history but only very gradually much later on. But did continents exist at all during the primitive period?

Richard Armstrong, then at Yale University, has claimed that continents have existed at their present size since the early days of the Earth and that continuing growth is a paradox, because continents are constantly destroyed, recycled into the mantle, and generated. The main result is that there is no growth in volume. Pieces of continents have different ages, but the total volume of the continents has been constant throughout geological time. Is such a theory true?

The Oldest Rocks in the World

The oldest rocks in the world are found in western Greenland. There is a complex of metamorphosed and folded rocks there that is in every way comparable to those that belong to recent periods. These relic formations (called Amîtsoq gneiss) near the town of Godthaab are 3.65 billion years old and were discovered and dated by Steve Moorbath's group at Oxford (Black et al., 1971). Nearby lies the Isua complex, which contains a conglomerate made of metamorphosed continental rocks. With Annie Michard, Joel Lancelot, and Steve Moorbath, my group at the University of Paris was able to determine that these rocks are 3.78 billion years old (Michard-Vitrac et al., 1977). At that point they were the oldest known continental terrestrial rocks.

But does this result really represent reality? Hasn't time completely

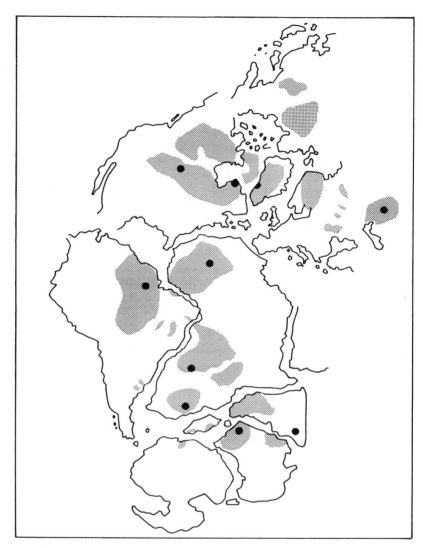

Figure 58 The continents 250 million years ago (Pangea) showing cratons older than 2.7 billion years (shaded areas) and rocks older than 3.5 billion years (dots).

erased what could have been our best document, rocks dating from Genesis? To answer this question, let us look at the sediments.

The sediments that are deposited at the bottom of the sea are products of continental erosion. Marine currents carry them away and mix them together. They thus constitute a real natural average of the continental run-off. Very old sediments—2.7, 3.5, and 3.8 billion years old—have been found (see Figure 58). They are a source of information about what had emerged on the terrestrial surface at the time they were deposited. If primitive continents much older than the period in which the sediments were deposited had existed, traces of them would be found in the isotopic composition of the sediments. The study of the isotopic composition of very ancient sediments tells us that, on the contrary, the continents that provided them were at that time hardly older than the sediments themselves (Hamilton et al., 1983). Some zircons (indestructible minerals that are found in all the sediments of the world) in Australia, however, have shown an age of 4.2 billion years. Since zircons are formed during granite formation, their age attests to the existence of pieces of continents 4.2 billion years ago, although no actual rocks from this period remain.

A complementary study of rocks from the mantle called komatiites, using neodymium isotopes as a tracer, has shown that large pieces of the continents were extracted from the mantle as early as 4 billion years ago. Thus a large volume has been destroyed. These results support Armstrong's model.

The Growth of Continents

Using all this information we can draw a curve of the evolution of continents. In the archaic period, prior to 4.2 billion years ago, continents were probably small if they existed at all and the Earth was entirely covered by ocean. About 4.2 billion years ago one or several continental nuclei started to emerge. From that time on the continental surface grew steadily. A period of rapid and important growth occurred between 4 and 3 billion years ago during which the foundations of North America, Brazil, Africa, India, Central Asia, Scandinavia, Scotland, and part of Central Europe were built. Then the rate of net continental growth decreased rapidly. For the last 500 million years the surface of the continents has hardly grown. The oceans cover two-thirds of the Earth's surface, but this percentage does not seem to be changing.

For the last 3.8 billion years, no net growth of the continents has occurred. Pieces of continents have been formed and extracted from the mantle, but at the same rate other pieces of continents have been destroyed. This destruction occurred through collisions between continents, which generated mountain belts like those in Tibet, and by erosion and subduction of the sediments generated. A kind of recycling machine has operated over the last 4 billion years, very active geologically but with an inefficient net result in terms of continental production. Was this period, like more recent times, governed by plate tectonics or was it very different, since convection was certainly more vigorous and the Earth hotter?

We have no answer yet.

The study of this period, which includes all the Precambrian terrains, is in full swing. It is of fundamental importance because of the new problems it poses and the new view of geologic history it provides: a history of long durations and irreversible evolutions. It is also extremely useful for our understanding of the geologic distribution of mineral deposits. Are South Africa, Australia, Canada, the United States, and the Siberian Shield rich in minerals and in ore deposits because the mantle ejected a large part of the metals it originally contained during the first 3 billion years of Earth's history? If so, the Sahara, Brazil, northern China, and all the Precambrian shields not yet explored offer a brilliant future for mining, since they must be well endowed with ores and minerals.

The Continental Extractor

Aluminum, potassium, and silicon, along with other useful metals, were extracted from the mantle over the course of geologic time to form the insubmersible crust of the Earth that is the continents. But how were they extracted? What was the extractor? What was the elevator? Volcanism, the only phenomenon that transfers materials from the depths to the surface, seems the necessary mechanism. But is it really?

Following Hutton and Lyell, we know that the "present is the key to the past." Let us examine present situations that seem likely. In subduction zones where the oceanic floor plunges into the mantle, for example, around the periphery of the Pacific Ocean, there is intense volcanic activity. It occurs in Japan and Chile as well as in

Indonesia and the Philippines. This volcanism is, however, very different from that which defines the midoceanic ridges or that which forms archipelagos like the Hawaiian islands in the middle of oceans. The latter, like lunar volcanism, ejects basalt, the most abundant rock in the solar system. The volcanism associated with subduction produces andesite (named for a common rock of the Andes mountains), which is rich in aluminum and potassium and is chemically very similar to the continental crust and the dominant granites that form it. The idea, already suggested, is that the continental crust is formed by the accumulation, directly above the subduction zones, of magma that is andesitic in composition. The continental crust is thus the daughter of the oceanic crust by andesitic volcanism. The process through which the mantle gives birth to a continent is associated with subduction and therefore takes place at the edges of continents. The centrifugal polarity of the geologic provinces, which has long been noted, thus appears to be a logical consequence of the location of the continental extractor near the continental borders, where the oceanic floor plunges into the mantle at subduction zones with their associated deep oceanic trenches. The formation of continents is a continuous process that makes use of the great cycles of plate tectonics. Yet several difficulties remain.

The majority of continental rocks are granites or metamorphosed ancient sediments. How are the andesites transformed into sediments, metamorphic rocks, and granites? Perhaps volcanic extraction is followed by erosion and deposition of sediments in the marine trenches, perhaps even by remelting, which finally produces the continental crust. Studies of trace elements (like barium, the rare earths, and uranium) in andesites show large differences between granitic and andesitic volcanics. At first they look similar, but when studied closely they appear completely different.

But the second problem is more severe. Since the major extraction of continents occurred very early in Earth's history, we need to know which processes operated at that time, and we do not. That it was probably not andesitic volcanism exactly similar to today's but very different is attested by several rock associations observed in old terrains. We have a long way to go before we understand everything. We have the techniques and we have the witnesses; now we must learn how to read the messages.

Having looked at the chronology of the development of the continental crust, let us turn toward the heart of the Earth, toward its core, to try to understand how such an "organ" was able to develop.

The Birth and Growth of the Core

The Earth's inner core is a solid surrounded by a liquid outer core. It consists of metallic iron alloyed with a little nickel (NiFe). The liquid outer core is moving at a rate of several kilometers per year, which is fast compared to the few centimeters of continental drift per year. The movement of a fluid conductor of electricity in a magnetic field engenders an electric current, which in its turn creates a field: this is the principle of the dynamo. Walter Elsasser (1939) and Edward Bullard (Bullard and Gellman, 1954) suggested that the core is a gigantic self-propelling dynamo. Thus, motion in the core is the source of the terrestrial magnetic field whose mysterious force orients the magnetic compass according to an axis practically identical to that of the terrestrial rotation (see Figure 59).

The structure of the core was determined through seismology, and variations in its magnetic field have allowed us to study its dynamics.

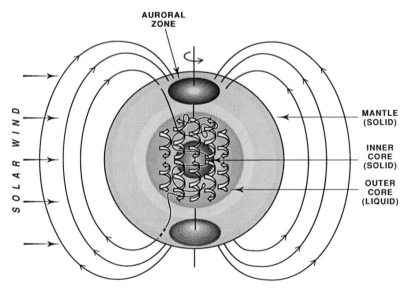

Figure 59 The core of the Earth and the liquid movement in the central core showing how it can generate a magnetic field.

The dynamics of the core, which were thought to be turbulent, chaotic, and disorganized, are perhaps extraordinarily simple, as Jean-Louis Le Mouel of the Institut de Physique du Globe in Paris has recently proposed. The circulation of the metallic fluid is polarized by the rapid rotation of the Earth, as is its meterologic circulation. The circulation has an equatorial symmetry dominated by two equatorial "spots": one located south of India where fluid from the interior of the core comes up, spreads out on the surface of the core, and falls back into a vast hole; and another off Peru, through which matter enters into the core. These dynamics were reconstructed through observations of variations in the terrestrial magnetic field over the last thirty years by magnetic observatories located around the world. Any movement implies a source of energy. What is the source in this case?

Iron is denser than silicates. The segregation of an iron core at the center of the Earth is a natural enough process to anyone who understands Newtonian physics. In a plastic medium the dense particles tend to concentrate in the center. This concentration corresponds to a variation in the potential energy of the system and therefore, since energy is conserved, to a dissipation of heat through friction. The formation of the core, like the accretion of the planets and the segregation of the inner core, creates heat. We can calculate that if the segregation of the core took 10 million years, the quantity of heat resulting from the process would have been sufficient to melt the entire Earth. If, on the other hand, the segregation took 4.5 billion years, the heat released had time to dissipate without dramatic consequences. As this calculation shows, the way in which the core was formed was not without a role in the history of the Earth. How can we discern the exact role of the differentiation of the core?

We have to go back to its chemical composition. As we saw in Chapter 2, seismic observations combined with laboratory experiments at high pressures allowed Francis Birch to assert that the principal component of the core is an alloy of iron and nickel. More precise measurements made more recently have altered this conclusion somewhat. The density estimated by seismologists is 10 percent lower than the density of the iron-nickel alloys measured in pressures of 1.5 to 3 megabars, that is, pressures corresponding to those in the core. This means that there must be light elements "dissolved" in the iron-nickel alloy that lower its density. What are these elements? As in the case of the iron meteorites we might suppose that the light compounds could be inclusions of silicates or iron sulphides. Ted Ringwood of the Australian National University thinks the light compound could be an iron oxide pulled along by and mixed with the metallic iron. High-pressure experiments using shock waves by Tom Ahrens of Caltech confirm the effect of light elements on the density without deciding between the various hypotheses.

Analysis of the mantle rocks supplies some parts of the answer. The siderophile elements (cobalt, osmium, rhenium, platinum, and gold) in the mantle are depleted compared to their concentration in meteorites. That makes sense when we recall that the core separated at the expense of the mantle. If we analyze the chalcophile elements (copper, sulfur, zinc, and lead) in the same way, we see that they too are depleted, which leads to the conclusion that they were incorporated into the sulfides in the core. Murthy and Hall of the University of Minnesota noticed that sulfur itself, although less volatile than nitrogen and the rare gases, is more depleted in the mantle. If it has not evaporated, where is it? Their answer: in the core (Murthy, Rama, and Hall, 1972).

Let us turn to the segregation process itself. Goldschmidt's image of the blast furnace implied a molten Earth, an immense bath in which little particles of solid iron fell to the bottom. It is difficult to accept this scenario today, because everything indicates that the Earth was never completely melted. Ringwood has clearly shown that the increase in the melting point of silicates with pressure makes such total melting impossible. So the differentiation of the core must have taken place in a permeable, porous medium in which the liquid loaded with iron was able to percolate through the solid silicate particles (see Figure 60). Such a phenomenon is impossible for pure iron or an iron-nickel alloy, since the melting temperature of these metals is greater than that of the silicates: when iron is melted, silicates also melt. The contradiction disappears if we assume that the liquid is an iron-iron sulfide (or iron oxide) mixture. Such a combination has a melting point that is much lower than that of either iron-nickel or silicates. We can therefore conclude that the core must contain 10 to 15 percent sulfur and oxide.

Now we can try to date the formation of the core by using lead isotopes. Lead is a chalcophile element. The formation of the core removed a large part of the lead from the mantle and changed its uranium/lead ratio. This event is reflected in the isotopic composition

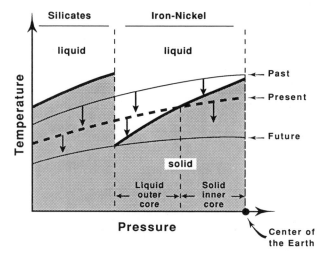

Figure 60 The explanation for a liquid core. The melting curve for silicates and iron-nickel is indicated by the solid line. As the Earth cools, the temperature curve goes down and the solid part of the core increases in volume. One can imagine a past time when the core was entirely liquid and a future time when it will be totally solid.

of lead originating from the mantle. Calculations show that the core formed between 4.45 and 4.3 billion years ago, but in a less sudden and violent way than Goldschmidt thought. The release of the heat engendered warmed the Earth but did not totally melt or vaporize it. The heat was sufficient to leave the core molten and thus to store the energy necessary to produce a magnetic field. This energy was given off progressively as the solid core grew. This is the delayed restitution of the energy of gravitation, the source of energy for movements in the core, terrestrial dynamics and, as a result, of the magnetic field.

The Mantle: Ancestor and Residue

Between the crust and the core is the mantle. Chemically speaking it is the source of both the crust and the core, and, as we shall see, of the atmosphere. When the core was extracted from it, the mantle lost most of its iron, and also the siderophile and chalcophile elements. During the continuous extraction of the continental crust, it lost part of its sialic elements, those that act like aluminum. And during the extraction of the atmosphere it lost some of its volatile elements.

Viewed dynamically, the mantle is the source of the surface movements called plate tectonics. The spreading of the ocean floor is not an autonomous phenomenon. It is only a surface or epidermal manifestation of broad currents called mantle convection that put most of the mantle into play and extend to depths of at least 700 kilometers. Although they are slow, with rates that are measured in centimeters or meters per year, on the scale of geologic time these currents are nevertheless curious and capricious. Their geometric characteristics and their rates vary continuously over time, that is, during the millions of years necessary to measure them. In fact, if we accelerated them by a factor of a few billion, they would appear very similar to those seen in a pot of water when it is vigorously heated from below. The mantle material is not water; it is the hard, solid crystalized mineral, olivine, which does not change its shape on a human time scale and acts like a real fluid only over the passage of millions of years. Geologic time gives properties to geologic materials that seem unbelievable but are nevertheless real.

Two complementary methods are used to study the mantle, its structure and its behavior. The first method is geophysical and maps the distribution of physical properties at different depths and in different geographic locations. Those physical parameters are density, wave propagation velocities, and electrical conductivity.

The second method consists of analyzing materials that come from the mantle. The mantle does not usually crop out on the surface. Its materials come to us as a result of two events. Sometimes tectonic processes bring pieces of it to the surface along the great faults. These form large massifs of peridotites whose structure and composition we can study, but these are limited in size. More conveniently, basaltic volcanoes bring to the surface incandescent lavas that were formed in the mantle. Basalts are very common on the Earth's surface, in the oceans as well as on the continents, and they offer a geographically representative range of samples. On the other hand, basalt is not a piece of the mantle; it is the most easily melted part of it, the part that is transported to the surface in the form of magma. Deducing the properties of the mantle from the composition of basalts is a little like trying to determine the ingredients in a recipe by tasting only its sauce! As in the case of the formation of the continental crust, the isotopic compositions of the elements linked with radioactive processes like strontium or lead are crucial. As I have said, isotopic composition does not change during melting, so the isotopic composition of strontium, neodymium, or lead in a basaltic magma is the same as that in the mantle from which it originated, which is certainly not the case for its total chemical composition (Gast, 1972). Their composition is evidence of their origin and the ages of their reservoirs.

By analyzing the isotopic composition of the basalts in the oceanic ridges, the oceanic islands, and the subduction zones we were able to discern differences in their origins and history, which led to an understanding of the structure and evolution of the mantle. Several groups, including those from Caltech, Cambridge, MIT, Mainz, and Paris have succeeded in such complex exercises both in technique and in interpretation. As a result of studying the mantle with isotopic tracers, we can now give a precise description of it. Let me summarize it briefly.

The mantle is divided into two layers. The upper one is 700 kilometers thick. Its vigorous motion drives plate movement. The continental crust was extracted from this layer over the course of geologic time. As a result, its chemical composition is depleted in aluminum and potassium but also in uranium and thorium, and therefore of a source of heat. The lower mantle lies beneath this layer. It is richer in aluminum and radioactive elements, and like the upper mantle it may contain convection currents. Recent studies suggest that its motion could be influenced by that of the core, but this is still only a hypothesis. A boundary layer at the bottom of this upper layer is the source of the hot spots that create oceanic islands like Hawaii, the Canary Islands, Iceland, the Azores, Réunion, and Tahiti. This boundary layer emits violent jets like those coming from the bottom of an overheated pot of water (slowed to the geologic time scale) (Allègre and Turcotte, 1986).

Energetics of the Earth System

The interior of the Earth is a dynamic system. Its movement necessitates energy. Where does it come from? By studying the Earth as a thermal machine we will be able to clarify its functioning. A good way of approaching the problem is to make an energy balance sheet for the Earth by measuring the heat flux at the surface.

The terrestrial heat flux is the amount of heat that escapes from the interior of the Earth each instant. Great progress has been made in measuring it, and we now have an average value (1.2 microcalories per square centimeter per second) and a map of its principal variations. How can this measured flux be explained? By remembering first of all that radioactivity gives off heat. Rutherford used this argument to refute Kelvin's argument that the Earth is young. Since then, great progress has been made by measuring on the one hand the amount of heat produced by the disintegration of uranium-235 and uranium-238, thorium-232 and potassium-40 and their disintegration periods, and on the other their abundances in the mantle and the crust. Yet when all the calculations are in, the total of these radioactive disintegrations accounts for only 50 percent of the present measured heat flux. Where does the other 50 percent come from? It might be a remnant of the energy that resulted from the accretion of the planet, the primitive differentiation of the globe, or the segregation of its core. But the mantle is convective and therefore well mixed; it conducts heat well. Storing such a large amount of heat in the mantle for such a long period is impossible. Therefore we must assume that the source of heat is the core, and that the heat is produced by the growth of the solid inner core at the expense of the outer core. Now we understand why the core could influence circulation in the lower mantle and, by extension, in the upper mantle, and therefore plate tectonics: it evacuates heat. Although this scenario requires much clarification, we are beginning to understand the Earth as a thermal machine. The upper mantle carries and propels plate movement. The energy for this, however, comes from below. Since it was depleted in radioactive elements when the continental crust was extracted, the upper mantle does not contain enough nuclear fuel to power itself. It is heated from the bottom and convects like an ordinary cooking pot.

Using our present understanding we can go back in time and "preview" the past. The production of heat in the interior of the globe must have been greater then because

1. Long-lived radioactive elements were more abundant and therefore more active 4.5 billion years ago; sources of radioactivity must have been ten times greater than today.
2. The effects of aluminum-26 and other short-lived radioelements may have been added to those of the long-period radioactive elements.

3. Heat created by the accretion of the Earth and the differentiation of the core added to the radioactive heat. Part of this accretion energy was expressed in the form of meteorite impacts and shock waves (remember that meteorite impacts and the accretion process are the same phenomenon).

Under these conditions the internal activity of the planet must certainly have been much more vigorous than it is today. Let us go further and try to imagine a more detailed, plausible scenario.

The Early Days

In its earliest infancy, the Earth was partially melted. The molten layer, 200 to 400 kilometers thick, appeared when the terrestrial radius became greater than 3,000 kilometers and stayed close to the surface. Pressure on the melting point of silicates solidified them as soon as they descended below a depth of 400 kilometers.

The ocean of magma was separated from the exterior by a thin crust, which was cracked, transported, constantly destroyed and constantly reformed. This terrestrial "ocean" was no doubt analogous in many ways to the lunar crust, but it did not form a thick crust of plagioclase like the moon, or rather, if it formed one, it destroyed it immediately. The reason is to be sought in the size of the planet. The much more massive Earth was richer in accretion heat, lost less of its energy through its surface because its surface to volume ratio was lower, and no doubt fed an extremely active convective system. This motion swallowed up the crusts as soon as they had formed, so the magmatic ocean has left few traces of their existence. On the surface the planet lost heat through radiation and cooled very quickly. In the interior the primary elements in the liquid NiFe-iron sulfite-oxide mixture began to percolate toward the center where the core was being formed.

About 4.3 billion years ago the core was nearly formed. The mantle was practically in the solid state. Very strong convective motion was still taking place at a rate of several meters per year, a hundred to a thousand times faster than today. The shape of the convection cells was probably very different than it is today. Using experimental analogies we can hypothesize that the cells were hexagonal, the ridges forming circles in the centers of the hexagons and the subduction zones running around the edges of each cell. This immense cycle was evident on the surface in the form of copious volcanism, both at the archaic ridges and at the subduction zones.

What was the nature of this volcanism? Was it basaltic like that of Mercury, the moon, or the present Earth? Or peridotitic, as the submarine lavas characteristic of terrains anterior to 1.5 billion years (called komatiites) seem to suggest? Or andesitic like the present-day subduction zones and like certain primitive belts associated with the komatiites? Or all three at the same time, as is probable? Much work remains to be done in determining its character, but there is evidence going back 3.5 billion years and methods of studying it are available. We will certainly know more soon.

What seems probable is that the primitive surface was populated by numerous volcanoes of various types. Earthquakes were no doubt restricted to the surface, because at that time the plates were very thin and must have dissolved quickly in the heat of the interior. The effects of a strong meteoritic bombardment must be added to this internal activity. The first pieces of continent began to form about 4.2 billion years ago on this Earth of fire and fury. Pulled along by convection currents and nourished by andesitic volcanism, the archaic continents began their incessant "ballet," while feeding on new materials and rapidly increasing their surface. Growth associated with destruction by erosion and drift associated with collisions were the rule until 500 million years ago. Since then, the continental rafts have led a calmer, more controlled existence, a regime well described in the plate tectonic model.

10

The Kingdom of Water

The geology of the Earth's surface is a product of the interactions—almost the antagonism—between the activity of the interior and that of the gaseous and liquid envelope, between the rocky medium and the fluid medium. It is dominated by the particular properties of a remarkable chemical compound: water. Present on Earth in the solid, liquid, and gaseous states, water is exceptionally reactive. It dissolves, transports, and precipitates many chemical compounds, constantly modifying the face of the Earth. Moreover, it is the essential component of a terrestrial peculiarity: life.

The Water Cycle

The functioning of the geological surface cycle has been well understood since Hutton's time. The principal actor in this cycle is water. Evaporated from the oceans, water vapor forms clouds, some of which are transported by wind over the continents. Condensation from the clouds provides the essential agent of continental erosion, rain. Precipitated onto the ground, the water trickles down to form brooklets, streams, and rivers, constituting what is called the *hydrographic network*. This immense polarized net channels the water toward a single receptacle, the ocean. Gravity dominates this entire step in the cycle, because water tends to minimize its potential energy by running from high altitudes toward the reference point that is sea level.

The rate at which a molecule of water passes through this cycle is not random but is a measure of the relative size of the various reservoirs. If we define residence time as the average time for a water molecule to pass through one of the three reservoirs—atmosphere, continent, and ocean—we see that the times are very different, as has been established by the use of short-period "natural" radioactive tracers like the tritium produced in atom bomb explosions and the carbon-14 produced by galactic cosmic rays in the upper atmosphere. A water molecule stays on average eleven days in the atmosphere, one hundred years on a continent, and forty thousand years in the ocean. This last figure shows the importance of the ocean as the principal reservoir of the hydrosphere, but also the rapidity of water transport on the continents. The flow of water over the continents is not harmless; all along its course water dissolves, infiltrates, penetrates, transforms, and corrodes the rocky materials it crosses. As it flows it transports dissolved salts and large and small solid particles downstream.

A vast chemical separation process takes place during this passage. Soluble ions such as calcium, sodium, potassium, and some magnesium are dissolved and transported. Insoluble ions such as aluminum, iron, and silicon stay where they are and form the thin, fertile skin of soil on which vegetation can grow. Sometimes soils are destroyed and transported mechanically during flooding. This can happen on mountain peaks during the first stage of soil development even before any vegetation has grown and just as chemical sorting has begun to operate. It can also happen when vegetation has grown solidly, which produces the red muds that are carried by the great tropic rivers in flood. The erosion of the continents thus results from two closely linked and interdependent processes, chemical erosion and mechanical erosion. Their respective interactions and efficiency depend on different factors.

Chemical erosion is more efficient when more water is present, rainfall is heavier, and temperature is higher. Higher temperatures increase the rates of chemical reactions, and for the dissolution reactions that interest us, a rough rule has been established: the rate doubles whenever the temperature increases by ten degrees. Tropical and equatorial zones are therefore particularly disposed toward chemical erosion. Greatly simplifying then, we can say that chemical erosion is a function of climate. In contrast, mechanical erosion depends mainly on altitude: the higher the relief, the faster the erosion. Mountains are worn down more quickly than hills, and hills more rapidly than plains. Numerous factors intervene locally to determine the shape of the relief, notably local geologic structure, the distribution of different kinds of rocks, and the geologic history of the area; but on the scale of entire continents statistical regularities are established and a few simple rules emerge. Continents are eroded until they reach an altitude of a few hundred meters above sea level, forming a peneplain through which rivers move in lazy meanders. At this stage, continental erosion stops.

Just as the rising of continents has global causes that are linked to plate tectonics, and more specifically, to collisions between continents, the formation of relief has a random distribution compared to the fixed distribution of erosion: the result is a diversity in terrestrial relief and the existence of erosion in very different stages corresponding to varying degrees of evolution.

The materials transported by the rivers end up in the ocean, where they are dispersed according to a well-defined system (see Figure 61). Those that are transported mechanically in the form of particles are deposited in zones along the coastlines: the large particles (sands and gravels) near the coast, the small particles (clays) offshore. The chemical elements that arrive in the sea in the form of dissolved salts go through a more complex process. Some, like sodium, remain mostly in solution, contributing to the salinity of sea water. Others, like potassium or magnesium, react with solid particles in suspension (clays) to produce new minerals or "feed" existing minerals. Still others, like calcium and silicon, enter the biologic cycle to produce shells or calcareous or siliceous tests (the hard, external coverings of plankton and invertebrates). When organisms die their tests—shells or exoskeletons—accumulate on the bottom.

The nature of the sediments near the coasts is a rather faithful reflection of erosion conditions on the continents. When the continents are mountainous, the sediments are sandy; when the continents are worn flat, the sediments are finer and contain a lot of shells. The nature of the offshore sediments more closely reflects conditions in the ocean: bottom depth, distance from shore, water temperature. When the bottom is shallow and the water temperature high, calcareous sediments dominate; when the depth is greater than 4,000 meters and the water is cold, fine red clays dominate. Thus, most of

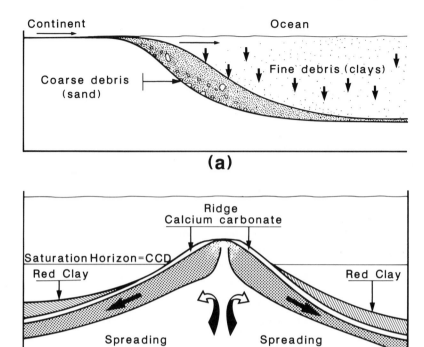

Figure 61 The sedimentation process in the ocean. a) Detrital sedimentation; b) Carbonate sedimentation.

the products of erosion are found in a direct or indirect way on the sea bottom in the form of sediments. The sediments accumulate horizontally, producing a series of superimposed layers that dry out to form those famous foliated archives called sedimentary strata. The strata associate as sedimentary series.

The sedimentary book must be read page by page, layer by layer, beginning at the bottom. Each page, each stratum, each rock has its own significance, which must be interpreted; but more important, the succession of messages deciphered in this way has a sequential meaning in geological language. After deciphering the letters comes the task of deciphering words and forming sentences according to a precise grammar. If a fine sand and then a clay are found on top of a coarse sand, we can conclude that the sea advanced farther and farther onto the coast, putting the local seafloor under deeper and deeper water. On the other hand, a limestone covered by a siliceous sand means that sediments from the coast suddenly riled up a warm, calm seacoast where limestone had been peacefully deposited. These messages, extremely simplified here, make sense only when the information from each stratum is put together and sequences appear, that is, when the words that are the strata are assembled into sentences. Reading the sedimentary series is like reading a book. The only difference is the size of the book and the nature of the language.

Fresh Water and Salt Water

Sea water has interesting properties, and the way in which its chemical composition is fixed reveals the functioning of that chemical factory, Earth. Simple common sense suggests that sea water is fresh water concentrated by evaporation. This would seem to explain the fact that sea water is saltier than fresh water. At the beginning of this century John Joly of Ireland attempted to calculate the age of the Earth using this idea.

But such a simple view is incorrect. Sea water is not just a more concentrated form of fresh water. According to its chemical composition, the most concentrated ion in sea water is the chlorine ion, followed by the sodium ion. In fresh water the most concentrated ion is bicarbonate, the chlorine ion is almost completely absent, and the sodium ion is rare, while those of calcium and potassium are abundant. The oceans are the natural receptacle of fresh waters but not an inert accumulator of them. A complex chemistry takes place in the ocean itself. Carbonate and calcium ions unite, notably in living things, to produce calcium carbonates, the essential constituents of limestone. Potassium, magnesium, and sodium ions are trapped in the clays. Of the ions contributed by continental run-off a certain number are subtracted from sea water.

But a series of other ions is injected into the ocean through the activity of submarine volcanoes, especially at the ridges. As was discovered several decades ago, sea water penetrates the oceanic crust as it is forming at the ridge crests and changes its composition, also producing submarine hydrothermal springs rich in metals and chlorides, trapping the sulfates in the form of sulfides, and performing a whole series of complex chemical transformations. If we look at the chemistry of sea water as a whole, we can say that the basic waters from the continents are neutralized by the acid waters from submarine volcanism. This was the prophetic description given by the great Swedish geochemist L. G. Sillen (1961), when the exact mechanisms of the volcanic contribution, exchanges between the ocean and the oceanic crust, and submarine hydrothermalism were still unknown.

The salinity of sea water is thus the result of a complex chemistry, but it exists in a kind of equilibrium state that in fact changes little over time. Contrary to what Joly thought, the salinity of sea water is not increasing but remains in a steady state.

Surface Water and Deep Water

Surface geology is dominated by the geologic role of water, but the realm of water is not limited to the surface. Water penetrates the terrestrial crust to a depth of almost fifteen kilometers. Near the surface in sedimentary terrains it digs the subterranean cavities that intrigued ancient peoples and that spelunkers explore. In the interior it percolates more discreetly but no less effectively.

Temperature increases with depth. Warm water has greater corrosive power than cold water, so deep waters dissolve and transport chemical elements with particular effectiveness. They form the majority of mineral deposits. Ancient peoples believed that the interior of the globe was rich in metals and sulfur and that it was thus the great dispenser of riches to man: the mantle raised useful metals to the surface in continental rocks, the transfer most often occurring without concentration into real deposits. Modern research has shown that except for chromium, platinum, and some nickel, most metals are concentrated by water in rocks already near the surface. Hot waters dissolve crystals; trace metals in the minerals pass into solution in the form of simple or complex ions, and alloyed to sulfide or hydroxide ions they produce complex ion compounds, the precursors of future ores. During the subterranean transport of these waters, metals separate out according to their affinities. A true selective filtration takes place.

To force a path for themselves, the hot subterranean waters flow through fractures and faults. The deposits they leave behind eventually seal them out, so that what is left are veins of metal or, more usually, veins of quartz or calcite. The formation of these ores and

Harmon Craig, Wallace Broecker, and the author at the American Geophysical Union meeting in Washington, D.C. Craig and Broecker were the pioneers in modern oceanography. They both received the Vetlesen Prize in 1988.

minerals has taken place over the entire course of geologic time and concentrated in certain places particular elements normally dispersed among the rocks. Today we have come to understand the chemical mechanisms that are the source of our riches by studying hot springs. They are found in all volcanic areas: Iceland, Italy, New Zealand, Japan, and the western United States. For the last twenty years the hot springs have been studied with the hope, realized in some places, of using their geothermal energy.

The first question was where did the hot water come from? In studying the isotopic composition of oxygen and hydrogen, Harmon Craig of the University of California at San Diego showed that the hot waters are rain water infiltrated and heated by the internal heat of the Earth in the top few kilometers of the crust. They do not come from the bowels of the Earth nor are they "youthful," as advertisements for companies that sell mineral water have claimed. Their isotopic composition, far from being uniform like that of the interior waters, mimics that of rain water, whose large geographical variations have been indexed and mapped (see Figure 62). This result was extended to the hot waters ejected by volcanoes, which also simply use rain water to discharge their formidable energy in explosions. Using Craig's results in a negative way we can say that the water

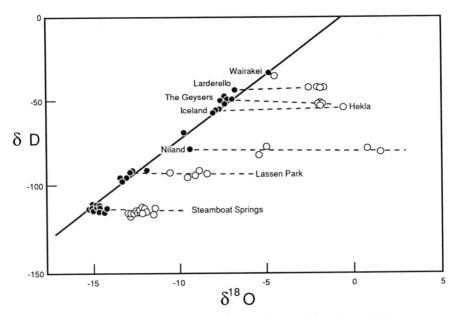

Figure 62 The isotopic composition of natural water. The diagonal line corresponds to rain water. The dotted horizontal lines represents geothermal waters. The vertical axis is the deuterium content, and the horizontal axis is the ^{18}O content as expressed in relative deviation from a standard.

found on the surface (salt water as well as fresh) is probably the result of the accumulation of all the hot springs as we know them today, not the gradual accumulation of water by present-day ejection.

The second question bears on the chemical composition of these hot springs. It has long been acknowledged that the hot springs of the volcanic countries located on continental crust, such as Japan and New Zealand, contain abundant dissolved metals, while the springs on volcanic islands, whose subbasement is purely *basaltic*, are depleted in them. This observation became a puzzle in 1978 when a Franco-American group discovered submarine hot springs in a *basaltic* subbasement that nevertheless vented thick black "smoke" rich in iron, copper, and zinc sulfide on the East Pacific Rise (CYAMEX, 1978). How could Iceland be poor in metals when the East Pacific Rise was ejecting deposits of them onto the sea floor?

The answer came in laboratory experiments, which showed that salt water, which contains sodium and chlorine ions, dissolves the metals dispersed in rocks a hundred times better than fresh water does. In the water of the Pacific Ocean the continental water that has percolated through salty sedimentary rocks has a mineralizing power because it has become salty, while the pure Icelandic rainwater does not. Gradually the mysterious veil over the mineral deposits was lifted.

Thermal springs, volcanic springs, the expulsion of water by volcanoes, fumaroles, metallic veins, and the deposition of giant gems in geodes are all different manifestations of the same phenomenon: the interaction of hot water and rocks. Today this is an active chapter of geology, which uses laboratory experiments, field observations, and calculations to penetrate the secrets of these fascinating mechanisms.

Sediment Cycles and the Establishment of Archives

The formation of sediments reflects the very effective processes of chemical separation on the surface of the globe. Erosion separates the soluble ions from the insoluble ones. Through the shift of differential sedimentation the various types of rock associations find a well-defined place of deposition. But conditions on the surface of the globe are changing. The shapes of coastlines vary, continents move, landscapes evolve. These changes are reflected in the lithological sequences of the great sedimentary series. Sedimentary series are made from solid continental rocks. What then becomes of these sediments?

Robert Garrels and Fred MacKenzie (1971), at that time at Northwestern University, considered the problem of the conservation of the sedimentary archives, which had received little attention. They found that sediments on the sea bottom have four possible futures:

- The vast majority of them, after being dried out, transform into sedimentary rocks, and, transplanted to the surfaces of continents, are in turn subject to erosion. They are destroyed, transported, and sedimented out again to produce new sediments. In the course of this enterprise they lose their identity and their age but not their isotopic composition. The global mass of the sediments act like cannibals: they feed themselves partly on their ancestors.

- A second rather large fraction is destroyed not by water but by internal heat. Buried in the depths during the process of mountain building, the ancient sediments are transformed by heat and pressure, metamorphosed, and sometimes even melted to produce granites. They then change their status: from sedimentary rocks they become metamorphic or even magmatic ones. In this way they contribute to continent building.
- A third part disappears into the mantle. Pulled along on the conveyor belt of the ocean floor, sediments follow it into the subduction zones, "contaminating" or "infecting" the mantle with pieces of the crust. In this way continental rocks are reinjected into the mantle and contribute to its chemical heterogeneity.
- In the fourth category are the survivors: the sedimentary series that, having risked all in the geologic lottery, traverse time, and arrive intact in the present. These survivors constitute the geologic archives. At what rate are they conserved? (See Figure 63.)

At present, 5 cubic kilometers of new sediments are formed per year, which corresponds to a mass of 4.10^{25} grams over 4 billion years. The present total of sedimentary rocks and sediments is only 2.10^{24} grams, twenty times less. So on average, one-twentieth of the possible documents have been preserved. Was the preservation faithful? In other words, do all sediments have an equal chance of being preserved? The answer is negative; the sedimentary archive is biased. If we want a faithful picture of ancient landscapes we have to learn to correct these systematic deviations.

An excellent census of all the sedimentary series in the world by rock type and age was made by A. B. Ronov of the University of Moscow. He found that the further back into the past he looked, the fewer examples he found for a given age. There are a great many sediments from the Tertiary era, fewer from the Paleozoic, and few from the Precambrian. This is the "life expectancy" effect we mentioned. Since sediments are susceptible to being destroyed in any period, their survival rate diminishes over time. When we examine the proportion of each type of rock that has survived, it becomes clear that the further back into the past we look, the more the proportion of limestone diminishes. This is to be expected, since it reflects the

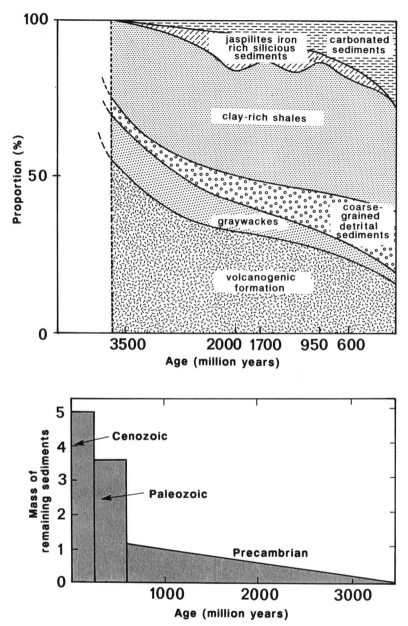

Figure 63 Conservation of sediments over the course of geological time. Top: Proportions of sediments of various ages; bottom: Abundances of sediments of various ages.

great vulnerability of limestone to alteration. In analyzing carbonaceous rocks we realize that the proportion of dolomite (calcium magnesium carbonate) increased in the past. But the clearest observation has to do with the much greater past abundance of sediments made of volcanic debris. Their resistance to erosion is somewhat lower than that of schist and sandstone, but they do not stop decreasing over time. This is no doubt a reflection of the amount of volcanic activity in the Precambrian period and of the relatively slight development of the continents at that time. The abundance of the magnesium ion, and therefore of dolomite, is related to the same phenomenon.

Sediments cover three quarters of the surface of the globe with a layer a few hundred meters and sometimes a few kilometers thick. This layer has probably existed since the earliest period; perhaps it was even thicker then. If the sedimentary epidermis is a geologic constant, however, its composition, distribution, and nature have evolved continuously. These variations are evidence of geologic evolution.

The Geologic Role of the Atmosphere

Let us use our global view to analyze the chemical significance of the erosion-sedimentation cycle. The alteration of the continents is caused by the action of water charged with carbonic acid. This acid is formed when the carbon dioxide in the air dissolves in water. The erosion of the continents therefore results in the "pumping" of the carbon dioxide out of the air. Once the carbon dioxide has been dissolved, it is bound up with bicarbonate and carbonate ions.

In contrast, the entire external cycle liberates the positive ions, or cations, such as sodium, potassium, and calcium, contained in rocks, transports them in their soluble form, and separates them from the insoluble cations of iron and aluminum.

In the sea the carbonate ion combines with calcium to produce calcium carbonate, which precipitates out. We can therefore picture the whole erosion-sedimentation cycle as an immense trap for the carbon dioxide in the atmosphere. Limestone is the great reservoir of terrestrial carbon dioxide gas, which illustrates the considerable chemical interaction between the atmosphere and surface geology. We mentioned it in connection with the water cycle; we also find it to be true for the carbon dioxide cycle.

What is the geologic role of the gaseous envelope called the atmosphere? The present terrestrial atmosphere consists of 80 percent nitrogen and almost 20 percent oxygen plus trace gases. Argon is the most abundant of these gases and reaches almost 1 percent. Water and carbon dioxide are minor but essential constituents. We know that the chemical composition of our atmosphere is very different from those of the other planets. Unlike the giant planets Jupiter and Saturn, whose atmospheres we know to consist of hydrogen and helium, our sister planets Venus and Mars have atmospheres that lack oxygen and in which nitrogen is very subordinate to the dominant gas, carbon dioxide. Why is the Earth so different in this respect from its sisters?

Before answering this fundamental question, let us return to the geologic role of the atmosphere. Its first task is to regulate the temperature and pressure conditions on the surface by reacting with solar radiation. The atmosphere is a buffer between the sun and the Earth. In effect, all the movement in the atmosphere, all of meterology and therefore the whole water cycle draw their energy from the sun's luminous radiation.

The Earth receives 263 kilocalories per square centimeter of energy in the form of radiation from the sun every year. Thirty-five percent of this flux is reflected back into space by the clouds in the atmosphere and the ice and snow of the polar caps. Part of the remaining 65 percent is absorbed while traveling through the atmosphere. The ultraviolet radiation is absorbed by a layer with a higher concentration of ozone at an altitude of 35 kilometers. Part of the infrared radiation is absorbed by water molecules and carbon dioxide. This means that the majority of the radiation that reaches the ground is in the visible spectrum. On the ground one part is reflected and another part is absorbed and reheated, either on the surface of the ocean or on a continent. The part that is reflected—20 percent on the continents and less than 2 percent on the oceans, which are black on satellite photos—does not even have the same spectral distribution as the incident part. Its spectrum is displaced toward the long wavelengths, that is, toward the infrared. This displacement is extremely important for the thermal equilibrium of the atmosphere. In effect, the carbon dioxide absorbs and therefore stops the infrared radiation. If the atmosphere contains a sufficient amount of carbon dioxide, the reflected radiation is trapped in the atmosphere and heats it intensely.

A large part of the absorbed radiation serves to evaporate water and therefore to foster the development of the meterological cycle, whose importance we have seen. Another part heats the surfaces of the oceans and continents and maintains the temperature at ground level as we know it.

The ground-level temperature varies with the latitude: since the amount of atmosphere traversed is greater toward the poles than at the equator, the proportion of the radiation that arrives at the surface is smaller at the poles. This establishes a climatic zonation at ground level that is spectacularly manifested by the two polar ice caps. But surface temperature is a very delicate parameter, which can fluctuate.

When the flux of solar radiation varies, the Earth's climate and its distribution vary. The variations take place with the seasons but also with more complex periodicities that are affected by variations in the revolution of the Earth around the sun, the spinning of the Earth on its axis, and variations in the activity of the sun itself. These are the cause of the famous glaciations that occurred periodically throughout geological time, as Louis Agassiz postulated in one of the hot geological controversies of the last century.

For more than twenty years it has been possible to study the climates of the Earth through isotopic analysis of fossil shells, as Harold Urey and Samuel Epstein, then at the University of Chicago, have shown. The $^{18}O/^{16}O$ ratio in the shells of living animals varies with the water temperature in which the animal lives but also with the $^{18}O/^{16}O$ content of the water. The microorganisms known as benthic foraminifera, which live mostly on the equatorial sea floor, grow a calcite shell whose $^{18}O/^{16}O$ ratio mimics that in sea water by a constant factor. Since polar caps have an $^{18}O/^{16}O$ ratio far lower than that of sea water, a large polar cap lowers the $^{18}O/^{16}O$ ratio of sea water, while a smaller polar cap raises it (see Figure 64). It is thus possible to learn the temperatures of ancient seas by measuring the isotopic composition of fossil shells. The "book" of the sedimentary series can then be used to reconstruct past climates, and such studies have been synthesized with great care, notably by John Imbrie of Brown University and Wallace Broecker of the Lamont-Doherty Geological Observatory.

During the last few million years of the Earth's history the climate has varied in a cyclical way, alternating glacial and interglacial periods. During the glacial periods the water in the polar caps was

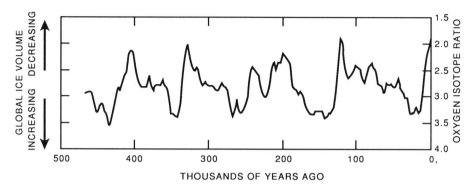

Figure 64 The variation in oxygen isotopes in seawater over the last million years, which reflects the variations in the average temperature of the surface as represented by the volume of ice in the ice cap.

subtracted from the ocean, causing marine regression, or retreat from land areas. During the interglacial periods melting of the ice led to marine transgressions, or extension over land areas.

Recent studies of climate variation through isotopic analysis of the oxygen in the polar caps and in foraminifera seem to confirm Milutin Milankovitch's theory, that is, the influence of variations in the solar cycle on the phenomena of celestrial mechanics. Since ancient geologic series are evidence of the existence of great marine transgressions (during the Tertiary period practically all of Africa was covered by water), we wondered whether the transgressions had been caused by gigantic warming trends: the melting of the present polar caps would cause sea level to rise 100 meters. It seems that this was not the case, but that periods of glaciation and warmer periods have succeeded each other over the whole course of geologic time. The study of long periods seems to show that a general warming tendency for the last 500 million years must be superimposed on these alterations, but this has not yet been proven.

Paleoclimatologic studies are very important to our understanding of how the temperature at the surface of the Earth has been maintained within reasonable limits over billions of years. A long period of freezing with an icy ocean and snowy continents would have caused Earth's climate to cool down forever; the white cover would have reflected sunlight so that the temperature stayed low. On the other hand, a hot episode would have vaporized the ocean and created torrid Venusian conditions that might have remained forever. But we have in fact escaped such extremes. The Earth's surface has been a well-controlled heating system. Boldly extrapolating to the earliest period of Earth's history, we can ask what the climate was like then. How did our present state evolve from this primitive condition? How have we escaped catastrophe?

To answer these questions we have to understand the crucial role of the ocean. For that we must return to our original question: what makes the composition of the terrestrial atmosphere so special? We have said that as a result of the erosion-sedimentation cycle carbon dioxide was stockpiled in the form of calcareous shells transformed into limestone on the ocean floor. Reasoning backward we can ask what the composition of the atmosphere would be if we destroyed the 10^{23} grams of limestone that exist on the terrestrial surface—for example, by heating. The atmosphere would then be dominated by carbon dioxide and would resemble the atmospheres of Venus and Mars: atmospheric pressure at ground level would be thirty times higher than it is at present. So the paradox disappears: the terrestrial limestone is responsible for the differences observed in the composition of the atmospheres of the three planets. Terrestrial carbon dioxide is trapped in the ocean as calcium carbonate. Thus we have explained one part of the mystery. What about oxygen, which is the second most abundant element in the Earth's atmosphere, although it is barely present on Mars and Venus?

Biogeology

The biosphere comprises all the living things existing on the Earth's surface, all the carbonaceous compounds that are living organisms. The mass of the biosphere (3.10^{17} grams) is negligible compared to that of the mantle (4.10^{27} grams), the core, or even the oceans. Yet it transforms and reproduces itself ceaselessly, by definition it is born and dies, so that if we calculate the total mass of living things that have existed in 4 billion years, it is greater than that of the continents. (Since the annual production of living matter is 6.10^{15} grams, that makes 2.4×10^{25} grams for 4 billion years. The mass of the continents is 1.4×10^{25} grams.)

The geologic role of the biosphere is considerable. We have already mentioned the fabrication of calcareous or siliceous shells, which form sediments (chalk is the best-known example). We have mentioned the role of vegetation, which retains the parts of rocks altered during erosion and produces cultivable soils from them. When some sudden phenomenon destroys this vegetation or it grows in coastal swampy zones easily covered by the sea, its remains mix with the sediments to produce coal deposits. In speaking of erosion we should also mention the role of a particular living being, Man, who can transport as much sand and gravel as a major river, and whose geologic role is becoming larger all the time.

What concerns us here, however, is different. It is the role played by life on the globe's surface and more particularly the influence of photosynthesis on the terrestrial atmosphere. Green plants, from the microscopic algae in the ocean to those superior vegetables called trees, absorb carbon dioxide and, using the carbon thus reduced, build living tissue. During the process they give off the excess oxygen. The transformation of inert carbon into living carbon is accomplished by a giant green molecule, chlorophyll. The process is called photosynthesis.

The excess oxygen in the atmosphere is therefore the result of *terrestrial life*. Without life there would be no oxygen. It is because there is no life on Mars and Venus that there is no oxygen in the atmospheres of these planets. The presence of oxygen has a secondary but essential effect: it allows animals that do not have the capacity to utilize carbon dioxide directly to draw their energy by eating planets and breathing oxygen. Oxygen, a product of life, is also the source of life. It is therefore an essential element of terrestrial evolution.

This realization immediately elicits a logical question: If life did not exist on Earth 4.55 billion years ago, there was no oxygen. The composition of the atmosphere has therefore evolved during geologic time. How did this come about?

The Ages and Origins of the Atmosphere and the Oceans

If the primary origin of the atmospheres of the giant planets is not in doubt, that of the Earth is more controversial. Did the atmosphere agglomerate around the solid Earth during the formation of the planet as Urey thought? Or did it result from the progressive degassing of the interior as W. W. Rubey suggested (this is called the "secondary origin" of the atmosphere)? We have only to observe a volcano to agree that it gives off an impressive amount of gas. An analysis of these gases shows that they consist of a mixture of water, nitrogen, carbon dioxide, sulfur dioxide, sulfurous gases, and rare gases. It is thought that some of them come from the mantle and are added to the atmosphere, continually increasing its volume, which leads to the idea that the atmosphere was formed through the degassing of the Earth's interior. This point of view fits perfectly with the idea that the primitive Earth had only a small amount of volatile elements on its surface. Only a small amount of nebular gas had been incorporated into the Earth through adsorption onto primitive dust grains.

If degassing from the interior did take place, it is important to know how the process developed. For a first scenario we can postulate that the formation of the atmosphere essentially took place at the beginning of the Earth's history and that subsequent contributions have been insignificant. The opposite scenario postulates that the atmosphere accumulated gradually over the course of geologic time at an approximately constant rate. Naturally we can imagine combinations of these two extreme scenarios: that a proportion resulted from the initial degassing and the rest accumulated over time. What is the reality?

Yet again it was by measuring the isotopic compositions of the isotopes formed by radioactivity that we elucidated this question, this time using the rare gases argon and xenon. Some of the isotopes of these gases are formed through radioactivity: argon-40 from the radioactive potassium-40, xenon-136 from uranium, and xenon-129, from iodine-129, a radioactive isotope that has disappeared today because its lifetime is short. Radiogenic isotopes are produced in the mantle. Since they are gases, they have a tendency to escape toward the atmosphere. There, deprived of contact with potassium or uranium, their isotopic composition is frozen and conserved. By measuring the isotopic composition of argon and xenon in the mantle and that in the atmosphere we can calculate the amount of degasification that took place. This process has been known in principle since 1950. The results, however, remained ambiguous for a long time because of inaccurate experimental results obtained on basaltic samples. The measurements were suspect and the models were problematical. Only recently by simultaneously using argon-40 and xenon-129 on well-chosen samples of fresh submarine basaltic glass could the problem be properly addressed and more precise calculations be made. Our calculations showed that 85 percent of the atmosphere was degassed in the first 10 million years of the Earth's history. The remaining 15 percent accumulated progressively, but with decreasing intensity. This was reflected in a degasification curve for the mantle that summarizes the phenomenon quite well.

The curve is important in explaining the evolution of the atmosphere but also in understanding the evolution of the internal dynamics of our planet. The degassing of the mantle is carried out principally by volcanism, the oceanic ridges being by far the most important contributor. The law of mantle degassing as a function of time is therefore also the law of seafloor spreading. Four billion years ago the rate of production of the oceanic crust was apparently twenty times faster than it is today. This is proof that convection in the mantle was also rapid, as we had concluded from calculating the amount of energy needed for degassing.

But another result of this calculation is extremely important. The atmosphere does not result from the degassing of the *entire* mantle. Only half contributes to its formation; but nearly 99 percent of this half has been degassed. This result affects our understanding of the archaic history of our planet in a surprising way. It confirms Ringwood's theory that our planet was never completely melted. In addition to the interesting deductions we can make about the behavior and the history of the mantle, this massive degassing indicates that we can consider the composition of the atmosphere as representative of the Earth as a whole, since whole outgassing means that the degassing took place without chemical fractionation.

The degassing curve obtained for the rare gases can be extended to include other gaseous compounds and water, and therefore the formation of the hydrosphere. The hydrosphere or the oceans appeared very early in Earth's history. This fits with Harmon Craig's finding that there is very little deep water in present-day volcanic gases. Where, then, did the water come from? Was it buried in the Earth in the form of water, or is it the result of a reaction of primitive hydrogen with the oxygen in silicates? A difficult question, but recently it has been argued that the water is original, that is, that it existed previous to the formation of the Earth. This hypothesis rests on various arguments. Water is found on other planets such as Venus and Mars. There is water in carbonaceous meteorites, and it has just been discovered in certain chondrites, where it is trapped in the form of fluid inclusions. Finally, the water molecule is observed in the cosmos. A small amount (1 percent) of water was therefore buried in the solid material and degassed as a liquid. A tentative hypothesis?

The Primitive Atmosphere and Ocean

What was the composition of the primitive atmosphere? Was it the same as it is today? Was the ocean as salty?

Since the primitive atmosphere resulted from the degassing of the interior of the globe, it is natural to try to determine what gases are imprisoned there today in order to learn its composition. This can be done by carefully studying the gases that escape from volcanoes and those that are trapped in minerals of deep origin. In both cases we find that, in addition to the principal gas, water, carbon dioxide is the major component, with nitrogen second. In short, the relative abundances of carbon dioxide and nitrogen are similar to those on Venus and Mars. What an interesting coincidence! Note that these deep gases contain only a few hydrogen compounds, such as methane and ammonia. They exist, but in low abundance.

For a long time, however, it was thought that these compounds were the essential components of the primitive atmosphere, as they are on Titan, a satellite of Saturn. If carbon and nitrogen are bonded with hydrogen, their ability to combine to produce large organic molecules, also rich in hydrogen, becomes even easier. On the other hand if, as is believed, carbon is bonded with oxygen and nitrogen remains free, carbon-hydrogen bonds must be formed to produce living material. This shows the importance of the problem for the origin of life. We will return to this question shortly.

Thermodynamics helped to solve this problem. If we calculate all possible reactions between gases and silicates, accounting for all the species present on the Earth's surface (including the carbon dioxide trapped in limestone) we can show that—except for water—the primitive atmosphere of the Earth was similar to that of Mars and Venus (see Figure 65). It was rich in carbon dioxide and nitrogen, while methane and ammonia were present in low abundance, as they are in the interior of the Earth today. That the Earth differs from Titan is the result of a property that we mentioned in connection with meteorites: the degree of oxidation or richness in oxygen. The Earth,

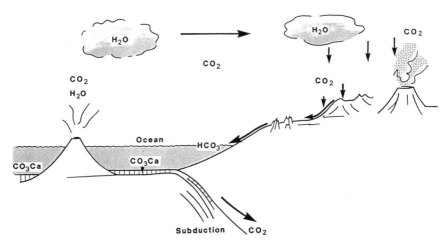

Figure 65 Early inorganic carbon dioxide geological cycle as it was on Earth before life occurred.

unlike certain meteorites and Titan, is rich enough in oxygen that part of its iron did not disappear into the core as an alloy but stayed in the mantle bonded to oxygen in silicates. Other gases, particularly very corrosive ones such as sulfur dioxide, were given off by archaic volcanoes. These reacted with water to produce sulfuric acid and hydrochloric acid. But of course water, the source of everything, dominated all the other gases in abundance. If all the gases were in the gaseous state at the same time, the pressure of the atmosphere at ground level must have been three hundred times what it is today. Under these conditions, if the temperature is not too high water passes into the liquid state, so the ocean formed immediately. The residual pressure of the atmosphere at ground level was then that of carbon dioxide: 50 atmospheres. At that distance from the sun an atmosphere so rich in carbon dioxide trapped solar radiation very efficiently. Moreover, the Earth's surface was hot because of internal heating, and volcanism was endemic and abundant. The atmosphere may have reached temperatures of 500° or 600°. If such temperatures have ever been reached, it was only briefly, because no liquid ocean could have been maintained at that temperature (water is limited to the gaseous state, no matter what the pressure, at 350°). More dramatically, water in the gaseous state would have been dissociated in the upper atmosphere by the effect of ultraviolet radiation, and hydrogen would have been lost to the planet. In such a scenario the Earth would have lost its precious water and would have been like Venus today: torrid temperature, no water, no life!

Therefore from the beginning a feedback mechanism must have played a role in preventing the accumulation of this large atmosphere and in decreasing its torrid heat. What we know of present conditions and reviewed at the beginning of the chapter indicates that the mechanism is to be found in the erosion-sedimentation cycle of the hydrosphere. Erosion acts like a trap for carbon dioxide, and sedimentation completes the process by fixing it in the form of limestone. For that to have happened, sea water must have already contained enough calcium to have been able to bond with the carbonates into limestone. Erosion of the rock must also have been very active, and the waters must have brought sodium, potassium, and calcium ions to the ocean, making it salty.

Let us imagine what this primitive ocean must have been like: degassed from the mantle at the same time as the other volatiles but more abundant than they, water arrived on the surface. The proto-hydrosphere was probably very acid because it had dissolved a certain amount of hydrochloric and sulfuric acids. It therefore attacked the volcanic rocks abundant at that time and rendered their calcium soluble. In other, more basic places the calcium and magnesium precipitated out in the form of limestone and dolomite, and the absorption of carbon dioxide began. The temperature was no doubt very high, but not too high: about 70°C seems likely. Such a temperature ruled out polar caps. On the other hand it activated chemical reactions, the alteration of rocks as well as the precipitation of limestones. The chemistry of the proto-ocean was more like that of the water found today near submarine hot springs than that of the present-day ocean. Gradually the ocean fixed the carbon dioxide and prevented atmospheric pressure from getting too high. This process continued for 10 million years. Water continued to accumulate on the surface and gradually formed the hydrosphere. Limestone was precipitated. The temperature fell. After 10 million years the Earth was covered by an ocean similar to the present-day ocean and by an atmosphere rich in nitrogen but containing not more than 10 percent carbon dioxide (see Figure 66).

How could such a system exist when there were no continents? What did the rain fall upon? What mountains could erosion work

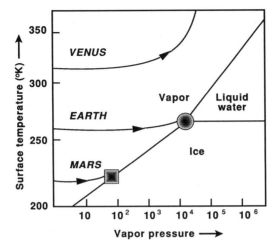

Figure 66 A phase diagram for water. The three lines with arrows are the trajectories followed by the atmospheres of Venus, Earth, and Mars.

upon? Where did the calcium necessary for the formation of limestone come from? As we shall see, the sun was at least as hot as it is today. It therefore created more severe thermal conditions in the atmosphere. The evaporation-transportation-precipitation cycle of the water molecules must have been more accelerated than the current one. And, in any case, since there were no continents, this cycle must have been much simpler, controlled only by general circulation and pole-equator transfer. Volcanoes that poked their heads above the primitive ocean, creating archipelagos or isolated islands, were immediately eroded. Sediments from the volcanoes began to form. This is confirmed by Ronov's studies of the relative abundances of the sediments as a function of time. The closer we go to the primitive period, the more important sediments of volcanic origin become.

Such a reconstruction seems logical and coherent, but at this point it rests only upon theoretical deductions. Dick Holland of Harvard assembled a whole series of data supporting the proposed scenario and filled in many of its gaps. First, we must note that the oldest geological formations in the world, those of Godthaab in Greenland, like those in Australia and Labrador, contain sedimentary series showing without ambiguity that 3.8 billion years ago an ocean and an erosion-sedimentation cycle similar to the one we know already existed. Studying these formations, some of which contain ancient, miraculously preserved continental soils and marine sediments, he was able to show that the minerals and the chemical composition of the deposits imply that carbon dioxide was then a hundred times more abundant in the atmosphere than it is today, that it did not achieve great pressures, that oxygen was absent and methane and ammonia were rare. The proposed scenario is therefore consistent with what the earliest rocks indicate.

The important element in this scenario is the absence of oxygen, which is confirmed by other geologic observations:

1. The sedimentary uranium deposits at Witwatersrand in South Africa, 3.4 billion years old. These contain uranium ores (uraninite) whose form attests that they were transported and sedimented as particles in a mechanical way. Uraninite is unstable and soluble in oxygenated water. Detritic pyrite (FeS_2), which is also unstable in oxidizing conditions, is present as well. Such deposits do not exist earlier in geologic history and are specific to ancient times.

2. Deposits of "chemical iron." On the terrestrial surface iron is soluble only in poorly oxygenated waters. In the oxidated state it immediately precipitates into ferric hydroxide, which is why it accumulates in tropical soils, turning them red. In the Precambrian (more than 2 billion years ago), iron ores were associated with siliceous formations whose origin by chemical precipitation is undoubted. Transport by fresh water in slightly acid conditions could only have occurred in the nonoxidated state. That implies an atmosphere poor in oxygen, determining in turn the same characteristic for fresh water, the agent of iron transport. When the water arrived in the "basic" marine milieu, it precipitated ferrous hydroxide at the same time as silica, from which come the deposits of "chemical" iron.

Everything seems to confirm the lack of oxygen in the primitive atmosphere. Since free oxygen exists on no other planet, we are led to connect its presence in the terrestrial atmosphere with chlorophyll assimilation and therefore with life. Holland tried to follow the evolution of the oxygen content by studying ancient sediments systematically. He was able to show that oxygen was still not very abundant 1.5 billion years ago and did not become really abundant until the last 500 million years. Plant life based on photosynthesis therefore developed very slowly.

Holland then naturally wondered about the chemical composition of the primitive hydrosphere. Following a detailed and systematic analysis of all the types of sediments found in periods anterior to 3 billion years, he concluded that the composition of sea water 3.5 billion years ago must have been identical to its composition today, except for its magnesium content, which must have been greater, and, of course, the nature of the dissolved gases that faithfully reflected the evolution of the atmosphere.

The Earth system, as complex as a living creature, with its feedback, regulation, and cycles that we can observe today, is the product of a long evolution and a long history. This history began when our planet formed and made itself conspicuous among its sisters. We have already reconstructed many steps in this history, but we have the means to progress much further in the years to come.

Clearly the archaic period was eventful, since the core of the Earth, a large part of the atmosphere, and the ocean were formed then. The convection currents in the mantle 4.5 billion years ago were a hundred times stronger than they are today and the sun was ten to twenty times brighter. All conditions were different, but nothing was complete. The slow maturation of time was necessary to produce the continents, the mantle, the atmosphere, and the ocean as we know them today.

Let us gather this history together to supply a chronological framework for the formation of the great terrestrial reservoirs (for without chronology there can be no true history). Using the dating of the formation of the continents, the core, and the atmosphere with the indications that we have about the sun, it should be possible to elaborate a more complete scenario. If, however, we want to situate the formation of our planet in the larger context of the accretion of meteorites and the formation of the other planets, one piece of information is lacking: the exact age of the Earth, a more precise age than that calculated by Patterson.

The xenon-129 content of the mantle and the atmosphere supplies us with this age. By comparing their isotopic composition with that of meteorites it can be shown that the Earth formed 50 million years *after* the meteorites. Remember that rubidium-strontium and uranium-lead dating fixed the formation age of the meteorites at 4.55

billion years. The Earth therefore formed 4.5 billion years ago at almost the same time as the metamorphism or volcanism of the small parent bodies of the meteorites was taking place. The sun had formed 50 million years previously and burned brighter than it does today.

The Earth's core formed very early; its average age is 4.42 billion years, which means that 4.3 billion years ago it had nearly reached its present size. The atmosphere and the ocean 4.45 billion years ago already possessed 85 percent of their present mass. They continued to enrich themselves and to change over the course of their history. As for the continents, they did not begin to appear until 4 billion years ago and formed very slowly over geologic time at the mercy of the process I have described. Thus we can complete our table of cosmic evolution.

It is clear that certain megastructures are the result of sudden archaic phenomena, while others bear the imprint of a long maturation. Earth's history should be conceived neither as a series of cycles nor as a series of catastrophes but within the framework of a long evolution, not omitting the distinguishing characteristic of our planet: life.

The Appearance of Life

The problem of when life appeared on Earth is probably the most fascinating one in contemporary science and also perhaps the most difficult. With a scientist's scruples I have long hesitated to approach it. Much is written on this subject and, as soon as one speaks, science and dream, illusion and proof quickly become confused. I hope that what follows will not add either to an already copious collection of nonsense or to the confusion. I will try to restrict myself to a few essential facts.

The Calendar

Living beings were already present on Earth 3.4 billion years ago. These were algae, unicellular beings that fabricated calcareous material that turned into limestone and stromatolites. Their fossil remains are found in the rock formations of Australia, South Africa, and Canada. Today the live origin of these fossils is not in doubt.

Life appeared on Earth in the billion years that followed its identification as a planet and evolved slowly, since the first multicellular fossil is found at the beginning of the Cambrian. It is an aquatic arthropod, a trilobite not more than 600 million years old.

It took life nearly 3 billion years to progress from single-celled to multicellular organisms. We known that a comparable interval was necessary to oxygenate our atmosphere and render it breathable.

The first mammals appeared about 200 million years ago and survived to the present, although their size and power did not seem to designate them the victors in their competition with the reptiles, then majestic and abundant. Man, the most finished product of evolution thus far, appears only 4 (perhaps 5) million years ago.

Nature, it seems, was a long time in finding her way!

A Decisive Experiment

In 1953 Stanley Miller (now at the University of California at San Diego) made a decisive experiment in Harold Urey's laboratory in Chicago, despite Urey's reluctance. In a vacuum flask he mixed methane, ammonia, and hydrogen and submitted them to electric shocks, thus fabricating a whole series of compounds typical of living material, up to amino acids.

Since then in numerous experiments using various original materials (including a mixture of hydrogen, carbon monoxide, water, and nitrogen with more or less methane and ammonia) and various experimental conditions (including shock waves, ultraviolet, and electrical discharges), complex molecules of the type found in living beings have been synthesized.

To summarize briefly, these thirty years of experiments established that it was not necessary to use compounds such as methane and ammonia to make organic molecules, and that oxygenated compounds such as carbon monoxide or carbon dioxide could work just as well. Yet no advance in molecular complexity that comes close to that in living beings has really been recorded, in spite of the cries of victory that were as boisterous as they were premature.

Clays and Replication

The discovery that all living things contain a common complex molecule called DNA and that the phenomenon of reproduction starts from it was an essential step in the understanding of life. DNA consists of a double helix of amino acids held together by a series of molecular "bars." In living things elementary reproduction takes place by cutting the bars and separating the two helices, with each separated helix finding the nutritive and organizational resources to reconstruct its other half. This is the mechanism of replication, considered the most important property of living organisms.

Recently, we have discovered phenomena of inorganic replication involving clays. Clays, typical products of natural alteration, consist of sheets of silicates separated by large ions and water molecules. When certain clays are treated with pure water the sheets detach from each other and produce a series of isolated flakes. Left to themselves the flakes remain isolated. But if the water is treated with dissolved salts, the flakes reproduce identical new sheets that are an exact "photocopy" of the old ones. A new injection of pure water separates the sheets again, and so forth.

We can imagine the alternation of phases of rain and of saline contributions in a coastal zone and understand how clays can reproduce themselves. From this some have deduced that organic molecules absorbed on the surface of such clays could have become "apprentices" in reproduction and then after some time acquired their own autonomy (Weiss, 1981). There is, however, a great distance between a phenomenon whose periodicity is guided from the exterior by climatic pulsations and the reproduction of living things, which seems to have its own internal clock.

Life in the Universe

Radioastronomers have discovered very complex organic molecules, such as alcohol, in the universe (Duley and Williams, 1984). Cosmochemists have discovered even more complex molecules in carbonaceous meteorites. All these molecules suggest those found in living things; some are actual amino acids.

Scientists, and not the least respected among them, have deduced from this that the synthesis of complex organic molecules is accomplished better in interstellar space than on planets, since in space radiation is much more intense. This suggests that primitive life was born in space. Brought to planets by the natural cosmic vehicles, meteorites and comets, these primitive beings would have contaminated all of them. Some planets would have had conditions that were favorable to these invaders; others would not. In the solar system, only the Earth . . .

Certainly scientific observations have established that synthesis of complex organic molecules is possible in abiotic conditions, in the midst of the great ocean of interstellar space. But is this not really the same message that is contained in Miller's experiments?

Life in the Hydrothermal Zones on the Ocean Floor

Over a decade ago, a Franco-American group diving in the Pacific ocean discovered submarine hot springs. They were hotter than 300°C. In their immediate environment, however, luxuriant life was discovered—in the very heart of these springs, bacteria seemed to flourish in temperatures of more than 250°C.

Reconstructing the extreme conditions that must have obtained on the surface of the globe, some see a picture of the development of primitive terrestrial life. Here again interpolation is rash. The discovery is important because it shows that living things can develop in conditions much more severe than we had previously imagined (above 100°C but at high pressure). It broadens the ecological conditions necessary for the appearance of life but does not solve its mystery.

In summarizing this research we will not add yet another scenario to the existing ones. What seems to be well established is that the primitive atmosphere from which or in which life appeared was rich in carbon dioxide. It was not a reducing atmosphere of methane and ammonia, as Urey and Miller thought, like that found on Titan. The first organisms must have invented the method of reducing carbon and creating living matter from carbon dioxide. In other words, they must have invented the equivalent of chlorophyll synthesis. It is clear that this "invention" was a decisive step in the development of life (Calvin, 1969).

After more than thirty years of research, we can conclude that today it is possible to conceive of how life began but that the fundamental mechanism is still not understood. The genesis of life is a physicochemical phenomenon conceivable with our methods of reasoning and our usual intellectual tools. We can reconstruct the chemical composition of the medium in which it appeared, but we have not taken the essential steps that would deliver the key to the phenomenon: the inorganic synthesis of DNA, for example, or the in vitro replication of DNA. As long as we have not made this step, scientific observations will continue to feed speculation on the frontiers of dream and mystery.

Cyclical Intellectual Behavior or Evolution?

Around 1830, with the triumph of the thesis of Hutton and Lyell, geology abandoned its research on the origin of the Earth and the way in which its great structures were formed. As Lyell had predicted, this distancing from cosmogony assured it peace with religious authorities.

For the last several decades, however—the moon landing was a decisive step—geology has again been preoccupied with these problems, with the successes and perspectives that we have outlined. Extrascientific problems seem to be beginning again.

In 1976 the Baptist church demanded that the state assembly of Minnesota forbid the schools to teach that the Earth is 4.5 billion years old. In 1979 in Arkansas in a now-famous law suit, a number of Protestant fundamentalist churches opposed the state on the subject of the age of the Earth. This time they requested that biblical chronology be taught along with that of the geological time scale. This case also revealed a new slant: certain witnesses who came before the bar were scientists, none a specialist in the field, obviously, who presented counter experiments "demonstrating" the inaccuracy of the law of radioactivity.

Over the last decade, the debate has continued.

Can the world of the mind, like that of matter, behave cyclically? Is the subject of Genesis really taboo for science? Let us hope that the exciting research being pursued today in quest of the origin of life and of the Earth can continue in the necessary serenity without prejudicing its results, without having them appear to anyone as offenses but rather as the natural progress of science. Science and religion are two separate aspects of human activity. They should stay separate. Nothing is to be gained by mixing the issues.

*

The star sun was born 4.55 billion years ago, and with it the first solid bodies in the solar system—preplanetary embryos of which chondrites are our precious relics. It wasn't until 50 million years later that the Earth as we know it today was born. Fifty million years to assemble hundreds of solid embryos, for the powerful solar wind of the infant sun to blow away the nebula's ambient gas. The birth of the Earth was no doubt simultaneous with that of Mercury, Mars, and Venus, but probably preceded the coalescence of the giant gaseous planets like Jupiter and Saturn and their satellites by 20 or 30 million years.

The Earth, which at first was bare, without atmosphere or ocean, its surface pocked with craters and pierced everywhere by glowing volcanoes, would evolve very quickly. In less than 30 million years it would secrete its iron core, vent its atmosphere, and form its ocean. At that time ocean covered almost the entire surface with a layer of water 2,500 meters thick, whose temperature was near the boiling point. In spite of the very brilliant, even glittering sun that lighted it, however, the newborn Earth avoided the diabolical trap of the greenhouse effect, which would have transformed it into Venus's twin and rendered it forever inhospitable to life. It quickly fixed the carbon dioxide in its atmosphere into submarine limestones and eroded the volcanic protuberances. External geology began (see Figure 67).

Then slowly, very slowly, it cooled off. Its interior cooled, its atmosphere and ocean cooled, volcanism quieted somewhat, and impacts became less frequent, but the water cycle continued its work, the carbon dioxide content of the atmosphere continued to decrease, acids began to be neutralized, and little by little the ocean became neutral, "habitable," welcoming.

The first continents to emerge from the primitive ocean were probably very quickly destroyed either by the infernal ballet of "archaic" seafloor spreading or by the impacts of gigantic meteorites. Perhaps one of these then snatched out a piece of the already differentiated Earth to create the moon.

Four billion years ago the situation had become calm enough for the continental embryos to grow and establish themselves definitively on the surface. The mobile but insubmersible and unsubductible continents grew very rapidly by extracting aluminum, silicon, and potassium from the mantle, a still very active mantle, danc-

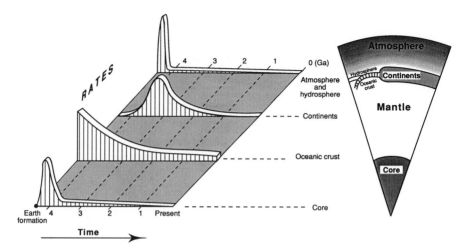

Figure 67 The rate of development of the Earth's reservoirs.

ing master of the surface ballet, whose chemical composition was depleted daily of the elements that nourished the other envelopes— core, atmosphere, and now continents. Continental growth continued over the course of geologic time until 500 million years ago but was counterbalanced by continental destruction through collision and erosion.

For the last 4 billion years the Earth's surface, divided into oceans and continents, has offered a spectacle animated by cyclical activity but also in constant transformation. The erosion-sedimentation cycle begun in the earliest period attempted to impose its periodic logic. Transformations and evolutions were linked to continental drift, to the size and the constantly decreasing rate at which the continents moved, the composition of the atmosphere, the invasions and retreats of the ocean, and the monotonous and cyclical evolution of climate.

On the Earth's epidermis at the interface between the solid ground and the fluid atmosphere and hydrosphere, life was born about 4 billion years ago. How? We still do not know, but we do know that during a slow evolution the number of species increased and the nature of the species changed. The reality of biological evolution is not in doubt, but its mechanisms and modalities are still obscure. Did the fall of meteorites or comets play a role in the process of natural selection? Was the cosmos perhaps a determining factor?

By studying stones and the messages inscribed in them—in the very heart of their atoms—we are beginning to discover a new history of planet Earth, to pierce the mysteries of its origin, of *our* origins. A new chapter in science is beginning, and we have perused a first version of it here.

We must remember, however, that it would not have been possible to reconstruct this whole beautiful history, this cosmic epoch, or to have situated it in its precise chronology if progress in laboratory experimentation and modern technology had not allowed us to measure the isotopic composition of atoms with extreme precision, to within one part in ten thousand. Within a few percent the isotopic composition of the planets seems to be homogeneous for most elements. But if we go into the precision of one part in a thousand, isotopic variations, and therefore questions, appear. At one part in ten thousand, answers begin to turn up and with them the challenge of making the extremely distant past live again, the past of origins begin to take shape. High resolution mass spectrometry was the key. Advanced technology was essential.

Pondering the cosmic scenarios we have evoked, we will understand nothing of the scientific adventure if we forget that all this does not emerge from the fertile imagination of theoreticians but from the patient, strenuous, often obscure and methodical daily work of thousands of researchers and technicians in scientific and technical laboratories around the world, permanently united in spirit. It is to them, my daily and steadfast fellow experimentalists, that I dedicate this book.

Appendix

References

Credits

Index

Appendix

The Atomic Structure of Matter

The *atom* is the basic unit in the structure of matter. Each bit of matter consists of millions of billions of atoms bonded together (an atom is about 10^{-8} centimeters across). But atoms are not all identical; there are various kinds. Each chemical element is characterized by a particular atom that has its own characteristics. Thus, an atom of hydrogen is different from a sulfur atom, an oxygen atom, or a uranium atom.

Atoms can bond together to produce complex chemical compounds. Two atoms of hydrogen can unite to produce a hydrogen *molecule,* but they can also unite with an oxygen atom to produce a compound that we all know: water. Its chemical formula, H-O-H, symbolizes the structure of the water molecule as we have described it, and it is the accumulation of billions of water molecules that forms, at normal temperatures, the liquid that we know.

But atoms themselves have an inner structure. They consist of a dense *nucleus,* which contains all the mass of the atom, and of *orbital electrons,* which are very light and orbit around the nucleus. The nucleus is like a central sun with the electrons as its peripheral planets. Electrical forces retain the electrons in their orbits. The electrons are negatively charged, while the nucleus is positively charged.

If we call the number of electrons (Z) and give each electron the unit charge $(-)$, the sum of the charges of all the electrons is $(-Z)$. To ensure a balance of electrical charge the nucleus thus carries a charge of $(+Z)$.

In fact, the number (Z) characterizes each element. In other words, each element is defined by the number of its peripheral electrons. The hydrogen atom (H) has a single peripheral electron (Z = 1), the helium atom (He) has two (Z = 2), the lithium atom (Li) has three (Z = 3), and so on. Oxygen has eight (Z = 8), and sulfur, sixteen (Z = 16). All ninety-two elements can be put in order by using (Z) as a parameter. The elements are arranged in this way in a table called the Periodic Table of the Elements devised by the Russian chemist Dmitri Mendeleyev.

Let us probe more deeply into atomic structure and examine that of the nucleus itself. It consists of two types of principal particles, the *neutron* and the *proton*. Protons and neutrons have almost the same mass (the same weight), but they differ in their electric charge. The proton is positive; the neutron, as its name indicates, is neutral. Since the nucleus has a positive charge (+Z) meant to compensate for that of the orbital electrons and each proton has a positive charge, the nucleus contains Z protons. In a given atom the number of protons is equal to the number of electrons: this is one of the fundamental rules of atomic structure.

But how do we determine the number of neutrons? No simple balance rule allows us to predict this number for a given element. Empirically, we see that the number of neutrons (N) is almost equal to the number of protons (Z) up to about 15, but when Z is greater than 15, the number of neutrons increases much faster than the number of protons. Thus, while the helium nucleus contains 2 protons and 2 neutrons, the lead nucleus contains 82 protons and 125 neutrons.

The explanation for the more rapid growth in the number of neutrons compared to the number of protons is found in very complex developments in particle physics, which call into play special forces like the weak interactions in a very complex formalism called quantum electrodynamics. A simplified explanation follows.

The nucleus is already a rather unstable structure, since it forces protons that all have the same positive charge to cohabitate. Because electrical particles of the same charge repulse each other, the nucleus would thus tend to explode. To avoid the repulsion the neutron plays the role of insulator and also of binding agent. The more the number of protons increases, the more chances they have to be adjacent and therefore to repulse each other, and the more necessary it is to increase the number of insulators in the form of neutrons.

The neutron therefore appears to be endowed with a property essential to the maintenance of nuclear stability but also with a certain liberty: if there must be a minimum of neutrons, nothing apparently limits the maximum number of them.

For a given Z, for a given number of protons and electrons Z—for the chemical element corresponding to the figure Z in Mendeleyev's table—several numbers of neutrons, N_1, $N_2 + 1$, $N_3 + 2$, and so on can exist. For the element Z it is thus conceivable to have three different nuclei: one with Z protons and N neutrons, one with Z protons and N + 2 neutrons, and a third with Z protons and N + 3 neutrons.

To illustrate the preceding with concrete examples, let us first consider the element hydrogen. It contains 1 electron. Its nucleus therefore contains 1 proton, but it can also contain 0 neutrons, 1 neutron, or 2 neutrons. The masses of these nuclei are 1, 2, and 3 respectively. If hydrogen is written symbolically as H, these three nuclei are called ^{1}H, ^{2}H, ^{3}H.

As we have said, neutrons and protons have the same mass. Nucleus (3) therefore contains 3 units of mass more than nucleus (1), and is therefore heavier. Nucleus (2) is intermediate in mass. Naturally these three nuclei having the same Z belong to atoms of the same element, and these three types of atoms have the same chemical properties although they differ from each other in mass.

Let us go on to a more complex atom, that of oxygen. It has 8 electrons and therefore 8 protons. But its nucleus can contain 8 neutrons, 9 neutrons, or 10 neutrons, producing three different oxygen nuclei of masses 16, 17, 18 (^{16}O, ^{17}O, ^{18}O). Various atoms of the same element that differ only in the number of neutrons are called *isotopes* of the element. ^{1}H, ^{2}H, ^{3}H are three isotopes of hydrogen, just as ^{16}O, ^{17}O, ^{18}O are isotopes of oxygen.

We may therefore wonder what isotope we are dealing with when we encounter an element in nature. Each natural element is a mixture of several isotopes, and each isotope constitutes a certain proportion of it. To measure the abundance of each isotope, we cannot use chemical procedures, since all the atoms belong to the same element. We must use physical properties that allow us to weigh each atom. We therefore need an *atomic balance.*

An atomic balance was actually invented by Francis Aston in 1914: it is the *mass spectrometer.* In this machine the atoms of a given element, previously charged electrically and accelerated in a vacuum to very high speeds, are separated by the action of a magnetic field. The heavier the particle, the more inertia it possesses and the less it is deflected by the magnetic field. The lighter it is, the more it is deflected. The charged atoms, the ions arriving in a magnetic field, behave like metal balls hitting a curved spring. If the ball, launched at great speed, is heavy, it diverts the spring. If it is light, it follows the spring. We can imagine a piece of apparatus based on this principle that would be capable of separating the balls according to their masses. Elements of this apparatus exist in all bars: they are pinball games. The ball is launched by a corkscrew spring and deflected by a curved spring. In the mass spectrometer the launcher is a source of ions, the deflector the magnetic field. The three isotopes of hydrogen separated in such a piece of apparatus have three different trajectories. The measurement of the electric current received at each impact point makes it possible to estimate the relative abundance of each isotope.

Since the invention of the mass spectrometer it has been theoretically possible to measure the isotopic composition of all the elements in Mendeleyev's table. In fact, however, the real situation is much more complex, because there are experimental difficulties with each element and measuring their isotopic compositions has necessitated as many original experiments as there are elements. But the principle was there, and based on previous work in mass spectrometry a catalogue of the isotopic composition of the different natural elements was established. This inventory showed that for each element there is only a small number of isotopes, which varies from one to ten. Some natural elements have a single nuclear structure, a single isotope. This is the case for sodium, fluorine, and manganese; others, such as tin and xenon, have nine or ten different isotopes.

Atoms can bond through their orbital electrons. When assemblages of atoms consist of only a few atoms, we speak of molecules. Thus, the union of two hydrogen atoms forms the *hydrogen molecule,* H_2. The union of two hydrogen atoms and one oxygen atom form the molecule of *water,* H_2O. Molecules contain two, ten, twenty, or hundreds of atoms. These chemical compounds can exist in different states: gas, liquid, or solid. The most common solids are crystals. Those structures formed by thousands of thousands of atoms have periodic structures. Sometimes these unions take place among thousands of atoms; then we speak of *macromolecules* or *crystals.*

Sometimes the formation of these compounds takes place according to a simple process. Certain atoms lose an electron; they become electronically charged and are called positive ions or *cations*. Other atoms gain electrons; they become negative ions or *anions*. In virtue of these laws of electrostatics, atoms of opposite charge attract and thus form compounds. Sometimes the formation of compounds takes place in a complex way through the common use of external electrons. This is what happens in organic compounds. Bonds in which the electrons are held in common are called *covalent* bonds (as opposed to *ionic* bonds).

References

Alfvén, H. 1954. *On the Origin of the Solar System.* New York: Oxford University Press.

Allègre, C. J. 1982. Chemical geodynamics. *Tectonophysics* 81:109–132.

Allègre, C. J. 1988. *The Behavior of the Earth.* Cambridge, Mass.: Harvard University Press.

Allègre, C. J., and D. Ben Othman. 1982. Nd-Sr isotopic relationship in granitoid rocks and continental crust development: A chemical approach to orogenesis. *Nature* 286: 335–342.

Allègre, C. J., and G. Michard. 1973. *Introduction à la Geochimie.* Paris: P.U.F.

Allègre, C. J., and D. L. Turcotte. 1986. Implications of a two-component marble-cake mantle. *Nature* 323: 123–127.

Allègre, C. J., S. R. Hart, and J. F. Minister. 1983. Chemical structure and evolution of the mantle and continents determined by inversion of Nd and Sr isotopic data. *Earth and Planetary Science Letters* 66: 191–213.

Allègre, C. J., T. Staudacher, P. Sarda, and M. Kurz. 1983. Constraints on evolution of Earth's mantle from rare gas systematics. *Nature* 303: 762–766.

Alvarez, W., F. Asaro, H. V. Michel, L. W. Alvarez. 1982. Iridium anomaly approximately synchronous with terminal Eocene extinctions. *Science* 216: 885–888.

Anders, E. 1971. Meteorites and the early solar system. *Annual Review of Astronomy and Astrophysics* 9: 1–34.

Anders E. 1977. Chemical compositions of the Moon, Earth, and encrite parent body. *Philosophical Transactions of the Royal Society of London,* ser. A, 285: 23–40.

Argand, E. 1922. *Congrès Géologique International.* Liège.

Aston, F. W. 1919. *Philosophical Magazine* 6: 38.

d'Aubuisson de Voisins, J-F. 1819. *Traité de Géognosie.* Vol. 2. Paris.

Audouze, J., and S. Vauclair. 1977. *L'Astrophysique nucléaire.* Paris: P.U.F.

Barrell, J. 1917. Rhythms and the measurements of geologic time. *Geological Society of America Bulletin* 28: 745–904.

Becker, R. H., and R. O. Pepin. 1984. The case for a Martian origin of the shergottites. *Earth and Planetary Science Letters* 69: 225–242.

Becquerel, H. 1896. *Comptes rendus Académie des Sciences* 122: 420–421.

Billingham, J. 1982. *Life in the Universe.* Cambridge, Mass.: MIT Press.

Birch, F. 1961. Composition of the earth's mantle. *Geophysical Journal* 4: 295–311.

Birch, F. 1965. Energetics of core formation. *Journal of Geophysical Research* 76: 6217–6221.

Black, L. P., N. H. Gale, S. Moorbath, R. J. Pankburst, and V. R. McGregor. 1971. Isotopic dating of very early Precambrian amphibolite facies gneisses from the Godthaab district, West Greenland. *Earth and Planetary Science Letters* 12: 245–259.

Bolt, B. 1982. *Inside the Earth.* San Francisco: Freeman.

Boltwood, B. B. 1907. Ultimate disintegration products of the radioactive elements—Uranium. *American Journal of Science* 23: 78–88.

Broecker, W. S. 1974. *Chemical Oceanography.* New York: Harcourt Brace Jovanovich.

Broecker, W. S., and J. Van Douk. 1970. Chemical oceanography. *Reviews of Geophysics and Space Physics* 8: 169–198.

Bullard, E. C., and H. Gellman. 1954. *Transactions of the Royal Society* 247: 213.

Buckland, W. 1820. *Vindicase—Geologicase or the Connection of Geology with Religion Explained.* Oxford.

Buffon, G-L. Leclerc, comte de. 1749–1783. *Histoire de la Terre.* Paris.

Burchfield, J. D. 1975. *Lord Kelvin and the Age of the Earth.* New York: Neale Watson.

Calvin, M. 1969. *Chemical Evolution.* New York: Oxford University Press.

Cameron, A. G. W. 1963. Formation of the solar nebula. *Icarus* 1: 339–342.

Cameron, A. G. W. 1970. The origin and evolution of the solar system. *Scientific American* 233: 32–41.

Carozzi, A. V. 1965. *Ohio Journal of Science* 65: 72–85.

Clayton, R. 1978. *Annual Review, Nuclear and Particle Sciences* 28: 501.

Clayton, R. N., L. Grossman, and T. Mayeda. 1973. A component of primitive nuclear composition in carbonaceous meteorites. *Science* 182: 485–487.

Clayton, R. N., N. Onuma, and T. K. Mayeda. 1976. A classification of meteorites based on oxygen isotopes. *Earth and Planetary Science Letters* 30: 10–18.

Conybeare, W. D., and W. Phillips. 1822. *Outlines of the Geology of England and Wales.* London: William Phillips.

Cook, A. H. 1970. The earth as a planet. *Nature* 226: 18–20.

Craig, H. 1963. *Conference on Nuclear Geology,* Torigiorgi, 17–53.

Craig, H., and J. E. Lupton. 1976. Primordial neon, helium, and hydrogen in ocean basalts. *Earth and Planetary Science Letters* 31: 369–385.

Cuvier, G. 1812. Preliminary discourse (Discours preliminaire) in *Recherches sur les ossements fossiles de quadrupèdes.*

Cuvier, G., and A. Brongniart. 1808. Essai sur la géographie minéralogique des environs de Paris. *Journal des Mines* 23: 421–458.

CYAMEX. 1978. *Naissance d'un Ocean.* C.N.E.X.O. Edition.

Darwin, C. 1859. *On the Origin of Species.* London: John Murray.

Dott, R. H., and R. L. Batten. 1981. *Evolution of the Earth.* New York: McGraw-Hill.

Duley, W. W., and D. A. Williams. 1984. *Interstellar Chemistry.* Orlando, Fla.: Academic Press.

Elsasser, W. M. 1939. *Physical Review* 60: 876–880.

Elsasser, W. M. 1963. Early history of the Earth. In *Earth Science and Meteorites,* ed. J. Geiss and E. D. Goldberg. New York: Wiley.

Emiliani, C. 1955. Pleistocene temperatures. *Journal of Geology* 63: 538–578.

Epstein, S. 1959. *Research in Geochemistry.* New York: Wiley.

Epstein, S., and T. Mayeda. 1953. Variation of O^{18} content of waters from natural sources. *Geochimica et Cosmochimica Acta* 4: 213–224.

Faul, H. 1978. A history of geologic time. *American Scientist* 66: 159.

Fowler, W. 1983. Nobel Lecture in Physics.

Fowler, W., and F. Hoyle. 1960. Nuclear cosmochronology. *Annals of Physics* 10: 280.

Ganapathy, R., and E. Anders. 1974. Bulk compositions of the moon and earth estimates from meteorites. *Proceedings of the Fifth Lunar Scientific Conference,* 1181–1206.

Garrels, R. M., and F. T. MacKenzie. 1971. *Evolution of Sedimentary Rocks.* New York: Norton.

Gast, P. W. 1960. Limitations on the composition of the upper mantle. *Journal of Geophysical Research* 65: 1287–1297.

Gast, P. W. 1972. The chemical composition of the Earth, the Moon, and chondritic meteorites. In *The Nature of the Solid Earth,* 19–40. New York: McGraw-Hill.

Geikic, A. 1897. *The Founders of Geology.* London: Macmillan.

Geiss, J., and H. Reeves. 1981. *Astronomy and Astrophysics,* 93–189.

Gillipsie, C. 1959. *Genesis and Geology.* New York: Harper and Row.

Goldschmidt, V. M. 1954. *Geochemistry.* Oxford: Clarendon Press.

Gray, C. M. and W. Compston. 1974. Excess ^{26}Mg in the Allende meteorite. *Nature* 251: 495–497.

Gray, C. M., D. A. Papanastassiou, and G. J. Wasserburg. 1973. Primitive $^{87}Sr/^{86}Sr$ in the Allende carbonaceous chondrite. *Eos* 54: 346.

Grossman, L. 1972. Condensation in the primitive solar nebula. *Geochimica et Cosmochimica Acta* 36: 597–619.

Grossman, L., and J. W. Larimer. 1974. Early chemical history of the solar system. *Reviews of Geophysics and Space Physics* 12: 71–101.

Gutenberg, B. 1959. *Physics of the Earth's Interior.* New York: Academic Press.

Hallam, A. 1983. *Great Geological Controversies.* New York: Oxford University Press.

Hamilton, P. J., R. K. O'Nions, B. Bridgwater, and A. Nutman. 1983. *Earth and Planetary Science Letters* 62: 263–272.

Head, J. W., C. A. Wood, and T. Mutch. 1976. Geological evolution of the terrestrial planets. *American Scientist* 65: 21–29.

Hess, H. H. 1962. *History of Ocean Basins in Petrological Studies.* New York: Geological Society of America.

Hohenberg, C. M., F. A. Podosek, and J. H. Reynolds. 1967. *Science* 156: 202.

Holland, H. D. 1984. *The Chemical Evolution of the Atmosphere and Ocean.* Princeton: Princeton University Press.

Holmes, A. 1911. *Proceedings of the Royal Society of London,* series 2, 85: 248.

Holmes, A. 1927. *The Age of the Earth.* London: Harper Brothers.

Holmes, A. 1945. *Principles of Physical Geology.* London: T. Nelson and Sons.

Hurley, P. M., H. Hughes, G. Faure, H. Fairbairn, and W. H. Pinson. 1962. Radiogenic strontium-87 model of continent formation. *Journal of Geophysical Research* 67: 5315–5334.

Hutton, J. [1795] 1959. *Theory of the Earth.* 2 vols. Reprint. Codicote, Herts.: Wheldon and Wesley.

Jacobs, J. A. 1975. *The Earth's Core.* New York: Academic Press.

Jacobsen, S. B., and G. J. Wasserburg. 1979. The mean age of mantle and crustal reservoirs. *Journal of Geophysical Research* 84: 7411–7427.

Jameson, R. [1808] 1976. *Wernerian Theory of the Neptunian Origin of Rocks: A Facsimile Reprint of Elements of Geognosy,* ed. George W. White. New York: Hafner Press.

Jeffery, P. M., and J. H. Reynolds. 1961. Origin of excess Xe^{129} in stone meteorites. *Journal of Geophysical Research* 66: 3582–3583.

Jeffreys, H. 1970. *The Earth.* 4th ed. New York: Cambridge University Press.

Kant, E. 1755. *Algemeine Naturgeschichte und Theories des Himmels.*

Kaula, W. M. 1968. *An Introduction to Planetary Physics.* New York: Wiley.

Kelvin, W. T., and J. J. Thomson. 1899. *Philosophical Magazine* 5: 47–66.

Kirwan, R. 1797. Examination of the supposed igneous origin of stony substances. *Transactions of the Royal Irish Academy.*

Kolodny, Y., J. K. Kerridge, and I. R. Kaplan. 1980. Deuterium in carbonaceous chondrites. *Earth and Planetary Science Letters* 46: 149–158.

Kurat, G. 1970. Zur Genese der Ca-Al-reichen Einschlusse im Chondriten von Lancé. *Earth and Planetary Science Letters* 9: 225–231.

Laplace, P. S. 1984. Exposition du système du monde. In *Corpus des Oeuvres de philosophie en langue française.* Paris: Fayard.

Larimer, J., and E. Anders. 1967. Chemical fractionations in meteorites 2: Abundance patterns and their interpretation. *Geochimica et Cosmochimica Acta* 31: 1239–1270.

Lee, T., D. A. Papanastassiou, and G. J. Wasserburg. 1977. *Astrophysical Journal,* 211.

Lehman, I. 1936. Bureau central séismologique international, série A. *Trauvaux Scientifiques* 14.

Lewis, J. S. 1973. Chemistry of the planets. *Annual Review of Physics and Chemistry* 24: 339–352.

Luck, J-M., J-L. Birck, and C. J. Allègre. 1980. ^{187}Re-^{187}Os systematics in meteorites: Early chronology of the solar system and age of the galaxy. *Nature* 283: 256–259.

Lyell, C. 1830. *The Principles of Geology.* Vol. 1. London: John Murray.

de Maillet, B. [1748] 1984. Telliamed. In *Corpus des Oeuvres de philosophie en langue française.* Paris: Fayard.

Marvin, U. B., J. A. Wood, and J. S. Dickey, Jr. 1970. Ca-Al rich phases in the Allende meteorite. *Earth and Planetary Science Letters* 35: 346–350.

Mendeleyev, D. I. 1871. *Loi periodique des elements chimiques.* French translation in *Le Moniteur Scientifique,* 21 March 1874, p. 653.

Mendeleyev, D. I. 1896. *Principles de chimie.* Paris.

Michard-Vitrac, A., J. Lancelot, C. J. Allègre, and S. Moorbath. 1977. U-Pb ages on single zircons from the early Precambrian rocks of West Greenland and the Minnesota River valley. *Earth and Planetary Science Letters* 35: 449–453.

Michel-Lévy, M. Christophe. 1968. Un chondre exceptionnel dans la météorite de Vigarano. *Bulletin Société Française Minéralogie Cristallographie* 91: 212–214.

Milankovitch, M. 1920. *Théorie mathématique des phénomènes thermiques produits par la radiation solaire.* Paris: Gauthier-Villars.

Miller, S., and E. Orgell. 1974. *The Origin of Life on Earth.* New York: Prentice Hall.

Millot, G. 1964. *Géologie des Argiles.* Paris: Masson.

Minster, J-F., J-L. Birck, and C. J. Allègre. 1982. Absolute age of formation of chondrites studied by the ^{87}Rb-^{87}Sr method. *Nature* 300: 414–419.

Morgan, W. J. 1968. Rises, trenches, great faults, and crustal rocks. *Journal of Geophysical Research* 73: 1959–1982.

Murchison, R. 1839. *The Silurian System.* 2 vols. London: Murray.

Murthy, V., H. T. Rama, and J. Hall. 1972. Composition and origin of the Earth's core. *Eos* 53: 602.

Mutch, T. A. 1973. *Geology of the Moon.* Rev. Ed. Princeton: Princeton University Press.

Mutch, T. A., et al. 1976. *The Geology of Mars.* Princeton: Princeton University Press.

Nier, A. O. 1938. *Journal of the American Chemical Society* 60: 1571.

Nier, A. O. 1939. *Physical Review* 55: 153.

Nier, A. O., R. W. Thompson, and B. F. Murphy. 1941. *Physical Review* 66: 112.

Oldham, R. D. 1900. *Philosophical Transactions of the Royal Society of London* 194: 135–174.

Oldham, R. D. 1906. *Quarterly Journal of the Geological Society* 62: 456–475.

Oliver, J. 1978. Exploration of the continental basement by seismic reflection profiling. *Nature* 275: 485–488.

O'Nions, R. K., N. M. Evensen, and P. J. Hamilton. 1979. *Journal of Geophysical Research* 84: 6091–6101.

Patterson, C. 1956. Age of meteorites and the earth. *Geochimica et Cosmochimica Acta* 10: 230–237.

Patterson, C. 1963. Characteristics of lead isotope evolution on a continental scale of the Earth. In *Isotopic and Cosmic Chemistry,* ed. H. Craig et al. New York: Humanities Press.

Playfair, J. [1802] 1956. *Illustrations of the Huttonian Theory of the Earth.* Facsimile reprint. Urbana, Ill.: University of Illinois Press.

Press, F., and R. Siever. 1974. *Earth.* San Francisco: Freeman.

Reeves, H. 1968. *Évolution stellaire et nucléosynthèse.* Gordon and Breach.

Reeves, H. 1982. *Patience dans l'Azur.* Paris: Éditions du Seuil.

Reynolds, J. H. 1960. *Physical Review Letters* 4: 8–10.

Ringwood, A. E. 1975. *Composition and Petrology of the Earth's Mantle.* New York: McGraw-Hill.

Ringwood, A. E. 1979. *Origin of the Earth and Moon.* New York: Springer-Verlag.

Ringwood, A. E. 1982. Phase transformations and the differentiation in subducted lithosphere: Implications for mantle dynamics, basalt petrogenesis, and crustal evolution. *Journal of Geology* 90: 611–643.

Ringwood, A. E., and A. Major. 1966. Synthesis of Mg_2SiO_4-Fe_2SiO_4 spinel solid solutions. *Earth and Planetary Science Letters* 1: 241–245.

Robert, F., and S. Epstein. 1982. The concentration and composition of hydrogen, carbon and nitrogen in carbonaceous meteorites. *Geochimica et Cosmochimica Acta* 46: 81–85.

Ronov, A. B. 1964. *Geochemistry* 8: 715–743.

Rubey, W. W. 1951. *Bull. Soc. Geol. Amer.* 62:1111–1147.

Runcorn, S. K., and K. M. Creer, eds. 1982. *The Earth's Core: Its Structure, Evolution, and Magnetic Field.* Royal Society Discussion Meeting, Jan. 27–28, 1982, Proceedings. London: Royal Society Scholium International.

Rutherford, E. 1906. *Radioactive Transformations.* New Haven: Yale University Press.

Safronov, U.S. 1969. *Evolution of the Protoplanetary Cloud and Formation of the Earth and Planets.* Moscow: Nantza.

Saint-Claire Deville, C. 1878. *Coup d'oeil historique sur la géologie et les travaux d'Élie de Beaumont.* Paris.

Sanz, H. G., and G. J. Wasserburg. 1969. Determination of an internal ^{87}Rb-^{87}Sr isochron for the olivenza chondrite. *Earth and Planetary Science Letters* 6: 335–345.

Schmidt, O. Y. 1944. *Meteoritic Theory of the Origin of the Earth and Planets.* U.R.S.S.: Dokl Akad. Nantz.

Sillen, L. G. 1961. The physical chemistry of sea water. In *Oceanography,* ed. M. Sears. Washington, D.C.: American Association for the Advancement of Science.

Smith, W. 1817. *Stratigraphical System of Organized Fossils.* London.

Staudacher, T., and C. J. Allègre. 1982. Terrestrial xenology. *Earth and Planetary Science Letters* 60: 389–406.

Steno, N. 1671. *Prodromos.* In S. Toulmin and J. Goodfield, *The Discovery of Time.* Chicago: University of Chicago Press.

Surkov, Y. A. 1977. *Proceedings of the Eighth Lunar Scientific Conference* 3: 2665–2685.

Tatsumoto, M., R. J. Knight, and C. J. Allègre. 1973. Time differences in the formation of meteorites as determined from the ratio of lead-207 to lead-206. *Science* 180: 1279–1283.

Taylor, S. R. 1982. *Planetary Science: A Lunar Perspective.* Houston: Lunar and Planetary Institute.

Toulmin, S., and J. Goodfield. 1965. *The Discovery of Time.* Chicago: University of Chicago Press.

Turekian, K. K. 1968. *Oceans.* New York: Prentice Hall.

Turekian, K. K., and S. P. Clark. 1969. Inhomogeneous accumulation of the Earth from the primitive solar nebula. *Earth and Planetary Sciences Letters* 6: 346–348.

Turner, G. 1977. *Philosophical Transactions of the Royal Society of London* 285: 97–103.

Urey, H. C. 1952. *The Planets, Their Origin and Development.* New Haven: Yale University Press.

Urey, H. C., and H. Craig. 1953. The composition of the stone meteorites and the origin of the meteorites. *Geochimica et Cosmochimica Acta* 4: 36–82.

Urey, H. C., H. A. Lowenstam, S. Epstein, and C. McKinney. 1951. *Bull. Soc. Geol. Amer.* 62: 399–416.

Uyeda, A. 1978. *The New View of the Earth: Moving Continents and Moving Oceans.* San Francisco: Freeman.

Van Schmus, W. R., and J. A. Wood. 1967. A chemical-petrologic classification for the chondrite meteorites. *Geochimica et Cosmochimica Acta* 31: 747–765.

Verhoogen, J. 1980. *Energetics of the Earth.* Washington, D.C.: National Academy Press.

Vernadsky, V. 1935. *La Géochimie.* Paris: Alcan.

Von Buch, L. 1802. *Geognostiche Beobachtungen auf Reisen durch Deutschland und Italian.* 2 vols. Berlin: Hande and Spener.

Wänke, H. 1981. *Proceedings of the Alphach Summer School,* ESA, July.

Wänke, H. 1983. *Origin of the Earth in the Solar System.* Vol. V/2. Landolt Bornstein.

Wasserburg, G. J., D. A. Papanastassiou, and T. Lee. 1979. Isotopic heterogeneities in the solar system. In *Les éléments et leurs isotopes dans l'univers.* Liège: University of Liège.

Wasserburg, G. J., D. A. Papanastassiou, F. Tera, and J. C. Huneke. 1977. *Philosophical Transactions of the Royal Society of London* 285: 7–22.

Wasson, J. 1974. *Meteorites.* Berlin: Springer-Verlag.

Wegener, A. [1929] 1966. *The Origin of Continents and Oceans.* New York: Dover.

Weiss, A. 1981. *Angewante chemie* 20: 850.

Weinberg, S. 1979. *Les Trois Premières Minutes de l'Univers.* Paris: Éditions du Seuil.

Wetherill, G. W. 1957. *Proceedings of the Second Lunar Scientific Conference,* 1539–1561.

Wetherill, G. W. 1976. *Proceedings of the Seventh Lunar Scientific Conference,* 3245–3257.

Windley, B. F. 1977. *The Evolving Continents.* New York: Wiley.

Wood, J. A. 1968. *Meteorites and the Origin of Planets.* New York: McGraw-Hill.

Wood, J. A. 1979. *The Solar System.* New York: Prentice Hall.

Wood, J. A. 1984. On the formation of meteoritic chondrules by aerodynamic drag heating in the solar nebula. *Earth and Planetary Science Letters* 70: 11–26.

Yoder, H. S. 1976. *Generation of Basaltic Magma.* Washington, D.C.: National Academy of Science.

Credits

Figure 2 After R. H. Dott and R. L. Batten, *Evolution of the Earth.* Copyright © 1981 by McGraw-Hill, Inc. Reprinted with permission.

Figure 3 From *Inside the Earth.* By Bruce A. Bolt. Copyright © 1982 by W. H. Freeman and Company. Reprinted with permission.

Figure 11 From R. H. Dott and R. L. Batten, *Evolution of the Earth.* Copyright © 1981 by McGraw-Hill, Inc. Reprinted with permission.

Figure 14 From Claude Allègre, *The Behavior of the Earth* (Cambridge, Mass.: Harvard University Press, 1988). Reprinted with permission.

Figure 39 From H. Reeves, *Patience dans L'azur* (Paris: Editions du Seuil, 1982). Reprinted with permission.

Figure 41 Adapted from W. S. Broecker, *Building a Habitable Planet* (Palisades, N.Y.: Lamont-Doherty Geological Observatory Press, 1974). Reprinted with permission.

Figure 53 (*top*) From Bryan Isacks, Jack Oliver, and Lynn R. Sykes, "Seismology and the New Global Tectonics," *Journal of Geophysical Research* 73 (1968). Copyright © 1968 by the American Geophysical Union.

Figure 53 (*bottom*) From *The New View of the Earth.* By Seiya Uyeda. Copyright © 1978 by W. H. Freeman and Company. Reprinted with permission.

Figure 59 From *Scientific American* 159 (September 1983). Copyright © 1983 by Scientific American Inc., George V. Kelvin. Reprinted with permission.

Figure 63 From Robert M. Garrels and Fred T. MacKenzie, *Evolution of Sedimentary Rocks* (New York: W. W. Norton, 1971).

Index